György Darvas

Hypersymmetry

Also of interest

Wilson Lines in Quantum Field Theory
Igor Olegovich Cherednikov/Tom Mertens/Frederik Van der Veken, 2020
ISBN 978-3-11-065092-1, e-ISBN (PDF) 978-3-11-065169-0,
e-ISBN (EPUB) 978-3-11-065103-4

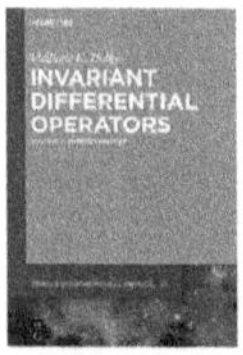

Invariant Differential Operators.
Volume 3: Supersymmetry
Vladimir K. Dobrev, 2018
ISBN 978-3-11-052663-9, e-ISBN (PDF) 978-3-11-052749-0,
e-ISBN (EPUB) 978-3-11-052669-1

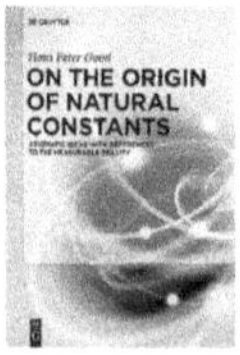

On the Origin of Natural Constants.
Axiomatic Ideas with References to the Measurable Reality
Hans Peter Good, 2018
ISBN 978-3-11-061028-4, e-ISBN (PDF) 978-3-11-061238-7,
e-ISBN (EPUB) 978-3-11-061048-2

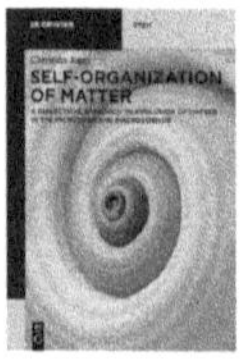

Self-organization of Matter.
A dialectical approach to evolution of matter in the microcosm and macrocosmos
Christian Jooss, 2020
ISBN 978-3-11-064419-7, e-ISBN (PDF) 978-3-11-064420-3,
e-ISBN (EPUB) 978-3-11-064431-9

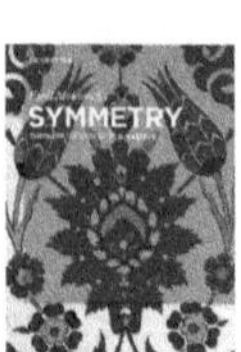

Symmetry.
Through the Eyes of Old Masters
Emil Makovicky, 2016
ISBN 978-3-11-041705-0, e-ISBN (PDF) 978-3-11-041714-2,
e-ISBN (EPUB) 978-3-11-041719-7

György Darvas

Hypersymmetry

Physics of the Isotopic Field-Charge Spin Conservation

DE GRUYTER

Author
Dr. György Darvas
Eötvös St. 29
Budapest 1067
Hungary
darvasg@caesar.elte.hu

ISBN 978-3-11-071317-6
e-ISBN (PDF) 978-3-11-071318-3
e-ISBN (EPUB) 978-3-11-071348-0

Library of Congress Control Number: 2020943675

Bibliographic information published by the Deutsche Nationalbibliothek
The Deutsche Nationalbibliothek lists this publication in the Deutsche Nationalbibliografie; detailed bibliographic data are available on the Internet at http://dnb.dnb.de.

Cover image: "Two boson exchange" by Kristóf Sarkady (8y) 2010
Typesetting: Integra Software Services Pvt. Ltd.
Printing and binding: CPI Books GmbH, Leck

www.degruyter.com

Contents

ANOTHER VERSION OF FACTS

1 **INTRODUCTION — 3**

Chapter I: **FIELD-CHARGES**

2 **MASS — 13**
2.1 Equivalence versus identity — **15**
2.2 Examples of the distinction between identity and equivalence — **17**
2.2.1 Physical approach — **17**
2.2.2 Mathematical approach — **19**
2.3 Equivalence does not mean identity — **20**
2.4 The equivalence principle — **21**
2.4.1 Equivalence of the masses of gravity and inertia — **21**
2.5 Transformation properties of gravitational and inertial masses — **22**
2.6 The role of masses in the stress–energy tensor of GTR — **25**
2.7 Conservation of mass — **26**
2.8 Some preliminary consequences of the distinction between masses of gravity and inertia — **28**
2.8.1 Where can one meet separated inertial and gravitational masses? — **30**

3 **ELECTRIC CHARGE — 31**
3.1 Distinction between electric charges — **31**
3.2 Sources of the electromagnetic field — **34**
3.3 Equivalence principle for electric charges — **36**
3.4 Transformation of the two types of electromagnetic charges — **36**
3.5 Some preliminary consequences of the distinction between field-charges of the electromagnetic field — **37**

Chapter II: **ISOTOPIC FIELD-CHARGES**

4 **ISOTOPIC FIELD-CHARGES (IFC) — 43**
4.1 Field sources in the standard model (SM) — **43**
4.2 Isotopic field-charges — **44**
4.2.1 Masses — **44**

4.2.2 Electric charges — 45
4.3 The identity-equivalence diversity on the example of the isotopic spin — 46
4.4 3 + 1 quantities in physics — 49

5 HYPERSYMMETRY (HySy) — 51
5.1 Matrix algebra for 3 + 1 parametric transformations — 51
5.1.1 Vector algebra and quaternion algebra — 51
5.1.2 Vector and quaternion algebras applied to physics — 54
5.2 The algebra of hypersymmetry (HySy) — 54
5.2.1 Introduction to the τ-algebra — 54
5.2.2 The τ-algebra — 55
5.2.3 Group properties of the τ-matrices — 56
5.2.4 Representation of the group composed by the τ-matrices — 59
5.2.5 The group of HySy — 60
5.3 Comparing the τ-algebra and the Dirac algebra — 64
5.3.1 The τ and the Dirac (γ) matrices — 64
5.3.2 Comparison of the algebras of the δ- and the τ-matrices — 65
5.3.3 The τ algebra beyond physics (matrix genetics) — 66
5.4 Summary of the τ algebra — 71

6 VELOCITY DEPENDENCE IN PHYSICS — 73
6.1 Velocity-dependent phenomena — 73
6.2 Velocity-dependent fields — 74
6.3 Velocity dependence in the light of conservation laws and symmetries — 75

Chapter III: **ISOTOPIC FIELD-CHARGE SPIN**

7 CONSERVATION LAWS AND HYPERSYMMETRY — 81
7.1 Preliminary assumptions — 81
7.2 Introduction to the mathematics of the two simultaneous Noether currents in HySy — 82
7.3 Noether's currents for gauge invariance localised in a velocity field — 84
7.4 Discussion of the mathematical results — 90
7.5 Physical considerations — 92

8 CONSERVATION OF THE ISOTOPIC FIELD-CHARGE SPIN (IFCS) — 93
8.1 First conserved quantity: conservation of the field-charge (ℸ) — **93**
8.2 Second conserved quantity: conservation of the isotopic field-charge spin (Δ) — **94**
8.3 Coupling of the two conserved quantities (ℸ and Δ) — **95**
8.4 Interpretation of the isotopic field-charge spin (Δ) — **96**

9 ISOTOPIC FIELD-CHARGES IN FUNDAMENTAL INTERACTIONS — 99
9.1 Isotopic field-charges in strong and electroweak interactions — **99**
9.2 Summary: field-charges in all the four fundamental interactions — **101**
9.3 Quanta of the D field — **103**

Chapter IV: **INTERACTION BETWEEN ISOTOPIC FIELD-CHARGES**

10 ISOTOPIC FIELD-CHARGES ($\daleth_V$ and $\daleth_T$) IN INTERACTION — 109
10.1 Mechanism of the interaction between IFC — **109**
10.1.1 Single particle's IFC states — **109**
10.1.1.1 The probabilistic model — **110**
10.1.1.2 The harmonic oscillator model — **111**
10.1.1.3 The flip-flop model — **111**
10.1.1.4 The intermediate particle model — **111**
10.1.2 The intermediate model of interaction between two particles — **112**
10.1.2.1 Symmetric or asymmetric interacting agents? — **114**
10.2 Interpretation of the IFCS conservation — **115**
10.2.1 On the roles of the masses once again — **116**
10.3 Mass of dions that mediate HySy transformations — **119**
10.3.1 Mass of the δ boson — **120**
10.3.2 Transformation in a coupled SM field and the D field – The origin of the mass of δ — **121**
10.3.3 Spontaneous breaking point of HySy — **129**
10.3.4 The mass of the mediating boson δ in light of the transformation of the D field — **133**
10.3.5 Conclusions on the dion mass and the transformation of the D field — **134**

11 IFC INTERACTIONS IN SM FIELDS — 135
11.1 Mechanism of commuting Δ — **135**
11.2 Hypersymmetry applied to gravitational interaction — **138**
11.2.1 Application of the τ algebra for the gravitational stress–energy tensor — **139**

11.2.2 Hypersymmetry of the gravitational equations — 143
11.2.3 The affine connection field — 145
11.2.4 The mechanism of the Δ exchange in gravitational interaction — 146
11.3 Hypersymmetry applied to electromagnetic interaction — 147
11.3.1 Isotopic electric charges in classical EM — 148
11.3.2 Isotopic electric charges in QED — 149
11.3.3 Isotopic electric charges in the presence of a kinetic gauge field — 155
11.3.4 Hypersymmetry of the extended Dirac equation — 157
11.3.5 Application of the HySy algebra for the extended Dirac equation — 159
11.3.6 Invariance of the extended Dirac equation — 161
11.3.7 Mechanism of the Δ exchange in electromagnetic interaction — 162
11.3.8 Modified Dirac equation in the presence of isotopic electric charges and a kinetic gauge field — 163
11.3.8.1 Coincidence with the classical Dirac equation in boundary case, when no kinetic field is present — 163
11.3.8.2 The magnetic and the electric moments — 163
11.3.8.3 The magneto-kinetic and electro-kinetic moments — 163
11.3.8.3.1 The magneto-kinetic moment — 165
11.3.8.3.2 The electro-kinetic moment — 165
11.3.9 The Hamiltonian and the Lagrangian of the electromagnetic interaction in the presence of isotopic gravitational and electric charges as well as a kinetic gauge field — 168
11.3.9.1 The full magnetic and electric moments — 168
11.3.9.2 The momentum in a kinetic field and the appearance of a virtual “coupling” spin — 170
11.3.10 The field tensors of the EM and the kinetic fields — 171
11.3.10.1 The field tensor of the EM field — 171
11.3.10.2 The field tensor of the kinetic field — 171
11.3.10.3 The curvature of the connection field — 172
11.3.11 The Lorentz force in the presence of a kinetic field — 173
11.3.12 The conserved currents and the conserved isotopic electric charge spin — 174
11.3.13 Quantisation — 175
11.3.14 Observation of a dion (a δ boson) — 177
11.3.15 Concluding remarks to Section 11.3 — 179
11.4 Mechanism of the Δ exchange in weak interactions — 181
11.5 Mechanism of the Δ exchange in strong interactions — 183

12 SUMMARY — 185
12.1 The birth and childhood of IFC hypersymmetry — 185
12.2 Summary of the findings in the HySy model — 188

12.2.1 SUSY and HySy — 188
12.2.2 What are those isotopic field-charges? — 188
12.2.3 Why we have not featured them in our physical equations? — 189
12.2.4 The question of localisation — 190
12.2.5 Interaction between the different components of a Hamiltonian? — 190
12.2.6 Why only opposite IFCS state particles can interact with each other? — 191
12.2.7 3+1 type quantities in physics — 192
12.2.8 The transformation group of HySy — 192
12.2.9 Velocity-dependent field? Velocity-dependent quantities in physics — 192
12.2.10 Conserved current – conserved quantity – mediating boson (dion) — 193
12.2.11 Properties of the isotopic field-charges and the δ bosons — 194
12.2.12 How much is the mass of a dion? — 194
12.2.13 Massive bosons in the D field? — 194
12.2.14 How many kinds of dion are sought? — 195
12.2.15 Why is HySy simpler than the SUSY model? — 195
12.2.16 In what features is (if at all) HySy more complicated than the SUSY model? — 195
12.2.17 Fermion–fermion and boson–boson pairs instead of fermion–boson pairs — 196
12.2.18 Do the two conserved Noether currents act together or separately? — 196
12.2.19 The SM and the HySy — 196
12.2.20 GUT and HySy — 197
12.2.21 Dark particles? — 197
12.2.22 Wave-corpuscle dualism — 197
12.2.23 Why does not the electron run away? — 198
12.2.24 Table of a few properties of isotopic field-charges — 198
12.3 Hypersymmetry and our picture of the physical world — 199
12.4 Closing remarks — 200

REFERENCES — 203

INDEX — 215

The physicist Leo Szilard *once announced*
to his friend Hans Bethe *that*
he was thinking of keeping a diary:
"I don't intend to publish. I am merely going to record
the facts for the information of God."
"Don't you think God knows the facts?"
Bethe *asked.*
"Yes," *said* Szilard,
"He knows the facts, but He does not know
this version of the facts."

ANOTHER VERSION OF FACTS

1 INTRODUCTION

We know from E.P. Wigner (1960) that physical theories are overdetermined in mathematical terms. This means, more mathematically correct theories can be formulated while only one is realised in nature. In mathematical terms, there are more correct, calculated results – in other words, theoretical "facts" – than in terms of physical reality. Experiments can decide which of them proves to be "THE" proper one.

Nature appears for us like a jigsaw puzzle consisting of asymmetric elements with a decorated surface. In geometrical terms, those elements can tightly tessellate a surface in various ways. However, only one of those tessellations will reconstruct a consistent physical picture painted on the tiles. So do mathematical theories attempting to describe the laws of nature. Several of them cover the surface tightly, but only one of them depicts nature correctly. All coverings represent different "facts", but only in experiments turns out which "fact" will be the correct depiction of nature. When that "fact" is found, one can say the sought law is discovered, the mathematically prepared theory is proven, "eureka".

One cannot predict in advance, which mathematically correct theory will substantiate the physical theory that describes nature correctly and which needs to be trashed. However, comparing with proven facts, checking their fitting to other well-proven theories, one can reduce the circle of candidate theories to a handful number. This preselection needs to investigate "other facts". "Facts" can be tested in experiments when the public is aware of those facts. For this purpose, the theoretical "facts" need to get published. First in parts in journal papers, conference lectures, university seminars, step by step, then summarised in book form. This book on hypersymmetry (HySy) looks at an extension of the standard model (SM), or a new physics (NP) beyond the SM, like many alternative theories of supersymmetry (SUSY) do. It argues for a few well-correlated "facts" tied in a bunch (like a bouquet). However, it presents "another version" of those known facts. The author hands this bunch to the reader.

-.-

Looking back into the history of physics, Newtonian mechanics described the physical world well for about two hundred years. At the end of the 19th century, it reached its limits. Several phenomena could no more be explained within its framework. At the beginning of the 20th century, there appeared two assumptions that transgressed the limits of classical physics. Both seemed "exotic" (or even absurd, in the eyes of others) at that time, but became productive and experimentally confirmed soon. One of them was the quantum hypothesis; the other was an upper limit of speed, identified with the velocity of light in vacuum. The first led to the quantisation of most physical quantities, the latter to the relativity theories. Both theories were effective because they did

https://doi.org/10.1515/9783110713183-001

not question the validity of classical physics at everyday distances and velocities, only extended it at that time extreme conditions. The quantum and relativity theories got their mature form through new mathematical tools, like operators and state functions interpreted in Hilbert spaces, group theory applied to physical invariances, as well as non-Euclidean geometries.

However, that time NP left open a few questions. It did not give a satisfactory answer to the problem of corpuscle-wave duality and could not exclude at all the existence of privileged (odd) reference frames (e.g. in which the amount of a conserved quantity is minimal). At the same time, two mathematical descriptions, the particle-model based matrix mechanics and the wave-model that was based on eigenvalues of wave functions proved to be equivalent descriptions of quantum mechanics.

The born new theories not only explained the experienced phenomena, for example, quanta of everyday matter and their behaviour in interactions, but they predicted new ones. The family of particles grew and all the predicted new family members were demonstrated in experiments among more and more extreme conditions, either in the cosmos or in laboratories. One should have made order among them.

Physicists always trusted in the simplicity of nature and found it in symmetry principles. Since Herodotus, whenever experience conflicted with belief in symmetry, that is, in the perfection of nature, scholars put the coin on symmetry even if it contradicted to what they observed. After nearly two millennia, Kepler was the first who gave up his strong belief in the absolute harmony (i.e. symmetry) of the universe when he accepted elliptical planetary orbits with the Sun in their odd focus. And yet, the priority given to symmetry survived and governed physical thinking against the complexity of the physical picture of the world. This approach dominates until now. The history of 20th-century physics witnessed a subsequent chain of discovering violations of symmetries, and then recovering symmetric order of the world at deeper levels, and so on. Symmetry principles have proved to be effective in the hands of physicists.

It is not surprising that the classification of particles and their interactions took place through symmetries – found at deeper levels. This led to a relatively simple classification of the existing and predicted particles and the unification of their interactions in the 1970s. This model – that clustered the particles in a symmetric table according to their characteristic properties, unified three of the four known fundamental physical interactions but gravity, and built in the classification also the particles mediating these interactions – is called SM.

The SM proved to be and is still effective. (We must mention that, similar to the early 20th-century theories, the SM could not solve satisfactorily the problem of the corpuscle-wave duality and the exclusion of the existence of privileged reference frames, either.) Nevertheless, to illustrate its effectiveness, twenty years after its formulation, by the mid-1990s, all particles but the Higgs boson (predicted in SM) have been experimentally demonstrated. However, one should not need to wait another two centuries to face the limits of the SM.

Experiments proved the SM's predictions, including the demonstration of the Higgs boson. It is a "powerful and brilliantly successful description underlying all the observed phenomena of nature that nonetheless leaves many loose ends" (Wilczek, 2018). SM does not need to get changed. However, as noticed, there appeared "loose ends" that demand its extension beyond its limits, since the 1990s. Although the SM was effective and explained the experienced phenomena in particle physics at the energies available for the physicist community, either in accelerators or through astronomical observations of that time, there were (mainly theoretically predicted) signs that at more exotic circumstances (e.g. at much higher energies) it reaches its limits. Almost parallel with the completion of the SM, the need for a NP (beyond the SM) was formulated. All agree that – similar to the relation of the relativistic and quantum theories to classical physics – the NP must be such an extension of the SM that does not question the latter's validity within its limits.

In the 1990s, when the need for such a NP was formulated, physicists knew certain constraints of the sought physics (that means, fitting to the SM), knew what questions they wanted to get answered, what shortages of the SM wanted to be solved, but they were poking about for something in darkness. We did not know at certainty the direction where to proceed. We agreed that the NP must be sought along with certain new symmetry, which is broken among the conditions where the SM prevails. However, as we have mentioned, we know from our experience that there are more mathematically correct models possible than what nature can realise. Nature realises only one of them. Since testing these models is expensive, preferably, one should exclude the less likely effective ones.

A theory that was available at that time was about the so-called SUSY. It proposed a possible model, although the model was based on exotic initial assumptions. One must put the question: how less exotic was the quantum hypothesis or the assumption on speed limit hundred years earlier? They are fully accepted now but were no less exotic that time. SUSY became popular soon. SUSY was based on string theories. String theories assumed additional, exotic space dimensions not observable in everyday life by our senses, but effective at extreme physical conditions. (Imagine, e.g. that a one-dimensional string vibrates in at least two dimensions and a two-dimensional membrane in three dimensions. What about the vibration of a three-dimensional brane? In this context, a fourth dimension seems no more too much exotic, cf., Kaluza–Klein theory. And so on to p-branes.) Disregarding details, based on superstring theories, SUSY proposed a new symmetry according to which all known fermions would have a bosonic brother, and all known bosons would have a fermionic brother. These brother particles were expected to be found at high energy collisions.

Gerard 't Hooft expressed his view on the NP after the SM: "What is generally expected is either a new *symmetry principle* or possibly a new regime with an altogether different set of *physical fields*" (see in Hooft, 2005, section 12; emphasised by me – G.D). The *isotopic field-charge spin* (IFCS) *conservation* (in the role of a new symmetry principle) and the (velocity-dependent) **D** *field*, being introduced in this book, are candidates (Darvas, 2009, 2011). Although the limits of the SM became apparent already in the 1990s, the expected NP is still to be waited for. The beliefs in finding SUSY particles were still strong in the mid-2000s, the Large Hadron Collider (LHC) experiments were still ahead – but after the start of the elaboration of the theory described in this book – when a few parallel workshops at the Conseil européen pour la recherche nucléaire, in Enlglish: European Organization for Nuclear Research (CERN) discussed possible theoretical candidate models beyond the SM to base a "new physics" in accordance with the fine-scale anomalies and symmetry breakings in high energy experiments.

Up to now, at 12–13 TeV energies in the LHC of the CERN, SUSY particles have not been found. As regards the chance of SUSY, "widely hailed as a great step forward in unifying our description of nature, has failed to materialize at the Large Hadron Collider despite a decade's worth of experimentation and anticipation" (Wilczek, 2018). Although there are still attempts to save the predictions of the SUSY, by applying more free parameters models, those are less reconcilable with the principle of simplicity (that nature seems to obey), and the hopes in its success are weakening. As regards a few other alternatives based on string theory, they have "failed to deliver concrete predictions, . . ., as have other, less heralded high-theory approaches" (Wilczek, 2018). Wilczek also confirmed the earlier prediction of Gerard 't Hooft set down on the role of symmetry. "Let me insert a few words in defence of beauty. Symmetry is at the core of the standard model and helped us to discover it. Modern physical cosmology also pivots on symmetry and simplicity, both in its general relativistic foundations and in its choice of initial conditions. . . . We need more beautiful ideas, not fewer."

Although several alternative theories have been formulated, the SUSY theory prevails among the activities of the majority of physicists. Most of those alternatives are based, alike SUSY, on string theories. Alternative theories predict particles to alternate the SUSY particles, a few of them have been studied, but less effort was invested in the search for them. Even the terminology – that calls these alternatives *exotic* particles – indicates that the attitude of the leading physicist circles to them.

Majority of physicists still believe to find SUSY particles (or at least a part of them) at higher energy collisions; the majority of physicists looking for the NP still believes in SUSY.

Simultaneously, alternatives to the SUSY have been revalorised. Their chances, including that of HySy, increased. In spite of the majority of the alternative theories, HySy did not start from a version of the string theories, it does not belong to those criticised by Wilczek, cited in earlier.

The philosopher Emile Chartier (1938) wrote: "Nothing is more dangerous than an idea when it is the only one we have". Up to now, there are about a dozen alternative theories of the SUSY. They all are *seeking symmetries of the new physics in different ways*. With the words of A. Szent-Györgyi (1985): "Discovery consists of looking at the same thing as everyone else and thinking something different." No one can predict which of those alternative ideas, theories, will ride first. Representatives of the alternative theories are in minority in their number. However, the validity of a physical theory cannot be decided by pottery ostracism. The hunt for "exotic" particles continues.

Alternative symmetric theories apply alternative "exotic assumptions". According to a saying attributed to A. Einstein, "If at first, the idea is not absurd, then there is no hope for it." If one allows accepting exotic dimensions in superstring theories for the SUSY, why not allow other, seemingly also exotic, seemingly "absurd", assumptions in the alternative theories?

-.-

This book treats fundamental physical interactions starting from two preliminary assumptions.

(*a*) Although the *mass of gravity* and *mass of inertia* are *equivalent* quantities in their measured values (at least, near to rest), they are *qualitatively not identical* physical entities. We will take into consideration this difference in our equations.

Later this "*equivalence is not identity*" *principle* is extended to sources of further fundamental interaction fields, other than gravity.

Where qualitatively different entities appear, one can assume that *physical interactions may occur between them.*

(*b*) Such interactions can take place in the presence of a *velocity-dependent field.*

These two assumptions do not contradict any known physical theory, known facts, while they allow "another" interpretation ("version) of facts" built in our explanations of physical experience.

The book first interprets the mentioned preliminary assumptions. It sketches, in main lines, a picture of fundamental physical fields influenced by the distinction between the two qualitative forms of the individual field-charges and interaction between them. The book describes the (theoretically predicted) existence of invariance between the two isotopic forms of the field-charges, and formulates certain consequences, in the light of the model, on the physical structure of matter. It describes a possible mechanism of their interaction, as it can follow from the mathematical derivations. This includes how these results can potentially change our approach to a few open questions of physics, including the properties and effects of a family of intermediate bosons predicted by the proven invariance between the assumed isotopic states of the individual field-charges.

The proposed conceptual framework and assumption on the interaction mechanism goes beyond the SM. As we showed, there is a common convincement that SM does not hold eternally alone and is not untranscendable; there should appear new, more precise theories that partially include the SM, and answer those questions that are left open by the SM. However, we do not certainly know-how, at least at present, although we have a few alternatives at our hands. The theory of HySy is one among these theories, an alternative to the most accepted, but experimentally not confirmed SUSY.

The theory of HySy, effective beyond the SM, and its predicted particles described in this book started to be elaborated in 2001, and many publications trace its road to get acknowledged since that. However, it has not received sweeping attention. It remained one among the approximately dozen alternative theories. The model presented in this book has been elaborated from quite different premises (see assumptions (*a*)–(*b*)) that radically discern it from the alternative models based on string theories. None of the two assumptions is quite new in physics, but their combination. I remind the earlier-cited words on the absurdity of ideas at first sight. Combination of the preliminary assumptions (*a*)–(*b*) looks like an absurd idea, nevertheless, they based the construction of a consistent theory. During the years, that theory proved to be mathematically correct. Experiments should confirm or reject its physical reality.

In defence of the apparently absurd idea presented in this book, I mention that it is based on the same facts like those considered in the SM, only on "another version" of them. It clusters the observations in another way. Unlike existing alternative theories, for example, the SUSY, which renders a new ("supersymmetric") brother to each particle, this model clusters the observed sources of fields in dizygotic twin pairs, regarding them as isotopic states of each other, and there is left "only" the twin siblings of the bosons mediating their interactions to be observed. It covers gravitational, electroweak, and strong interactions. In contrast to the SUSY, which renders fermion–boson pairs as new-born siblings to each other, the IFCS assumption, proposed in the present work, renders fermion–fermion and boson–boson twins to each other.

The IFCS assumption does not predict new fermions; the twin siblings of fermions originate in splitting the existing ones. The preliminary assumptions are not quite new. Known "facts" are interpreted in "another version". Moreover, these preliminary assumptions are combined in a way that has not been applied in physics. The *field-charge siblings*, like gravitational and inertial masses, as well as Coulomb and Lorentz charges, were at our hands for centuries, but one has not distinguished them in the usual physical equations. Why? Because there did not appear a necessity for it at everyday laboratory conditions. *Velocity-dependent fields* were known for decades but were exhaustively applied only for less fundamental phenomena. Thus, the novelty lies in their combination. The feasibility of the interaction between field-charge siblings is a consequence of the two latter "versions of facts" (i.e. field-charge siblings and velocity field).

The idea is no more "absurd" than the assumption of SUSY particles was in the 1980s (or asymptotic freedom in the 1970s, quantum hypothesis in 1900, and so on aback). I know from my personal experience that most of my physicist colleagues, who invested decades during their career into finding SUSY particles oppose this model since its birth. Many that believe in other "exotic" particles misdoubt. I understand them. However, no pottery ostracism, only experiments can decide among the alternative models. The outcome does not depend only on how much was staked on this or the other model; let us remember the fate of the SUSY, how much labour force and money were invested in confirming it. At the same time, I am indebted to those colleagues who stood with me, even in the most hectic periods, when neither I nor they were convinced in the successful outcome of the elaboration of the HySy theory. Now, you are keeping it in your hands. I hope this patience was not presumptuous.

Discovery consists of
looking at the same thing
as everyone else and
thinking something different.
Albert Szent-Györgyi

Chapter I: **FIELD-CHARGES**

2 MASS

Mass is a physical quantity familiar to the reader. It plays an odd role in physics. Therefore, we start the discussion with mass.

The difference between gravitational and inertial masses was known for over three hundred years. Why did not one distinguish them in the equations of physics? There were different reasons for it.

First, according to a simplified formulation of the second law of Newton, the mass can be defined as the ratio between a force that affects it and its acceleration caused by that force. This definition does not take into consideration whether the force in question originates from the gravity of another object or it is an inertial force. The so defined mass could be measured by the same unit in both cases.

Second, the proportion between the measures of the two kinds of masses was fixed to "1". This made the appearance like the two masses were identical, and at the available test procedures, the results seemed satisfactory for centuries.

Third, the so-called equivalence principle was formulated. The quantitative equivalence between the gravitational and inertial masses of the same body (at least at rest) was measured accurately first by R. Eötvös (and his colleagues, 1910, 1922) in 1906–1909. The experimental proof of their equal value made available for A. Einstein to formulate the equivalence principle in the early 1910s. When Einstein (1915, 1916) formulated the general theory of gravity, known as the general theory of relativity (GTR), he really considered that the equivalence principle meant identity between the equivalent value masses. Although he changed his mind in a few years and formulated more precisely, the so-called "weak formulation of the equivalence principle" inspired many textbook writers to identify them, even up to now. This was pragmatically correct, but theoretically not justified. The problem became acute when technological development made possible to observe experimentally high energy interactions between massive objects. We will discuss the difference between equivalence and identity in Section 2.3, and the qualitative difference between the two kinds of masses at relativistic velocities in Section 2.5.

Preliminary, we give a simple, school textbook instance to exemplify the difference between the two masses. When a ball drops, its potential energy (denoted by mgh, where m is the mass of the ball, g is the gravitational acceleration at the given place, and h is its height over the floor where it will fall) transforms into kinetic energy ($mv^2/2$, where m is its mass again, and v denotes its velocity when it reaches the floor). Since the conservation of energy is a very strong principle in physics, the values of the initial and final energies must be equal:

$$mgh = mv^2/2$$

Let us note that the mass appearing in the left side of the equation is the measure of gravity proportional to the force ($m_g g$) by which the Earth attracts the ball; while

https://doi.org/10.1515/9783110713183-002

the right side of the equation is a kinetic expression, and the mass appearing in it is proportional to the inertia ($m_i v$) of the ball. Instead of $gh=v^2/2$, we need to write

$$gh = (m_i/m_g)v^2/2$$

R. Eötvös (and his colleagues, Pekár, D. and Fekete, E.) measured the fraction m_i/m_g and found that it is equal to 1 at high precision. However, he performed the measures in rest. The two masses are quantitatively equal until the speed of the ball is not relativistically high. The Eötvös (1910, 1922) experiments do not allow us to be certain if the two masses will behave in the same way at velocities too far from the rest. The Lorentz transformation intimates to put the question of whether they will change their behaviour. Maybe yes, maybe not, but we need to obtain evidence. It was only in 3–4 years, when M. von Laue and then more intensively G. Mie discussed the difference in the transformation at an increasing velocity of the two kinds of masses (Section 2.5). In course of the way towards those evidences, we must investigate the transformations of the gravitational and inertial masses (Section 2.5) and, depending on the results, reconsider the conservation rules of the two masses (Section 2.7).

There were different attitudes to these questions during the last hundred years. We will discuss them in Section 2.5. Nevertheless, at this stage, we have no right to denote the two masses by the same character. According to the traditional signage in physics, the potential energy was denoted by V, and the kinetic energy by T. Since the gravitational mass appears in the potential (scalar) part of a Hamiltonian (V), we will denote it by $m_{\text{gravity}}=m_V$. While the inertial mass is the source of the kinetic (vector) part of a Hamiltonian (T), we will denote it by $m_{\text{inertia}}=m_T$ in the following. So, our equation looks like

$$m_V gh = m_T v^2/2\ .$$

Fourth, there arose a problem about the relation of the two kinds of masses. One easily accepted that different kinds of energy may transform into each other and conserve the quantity of energy. The equations of such transformations show that different physical quantities appear in their two sides. In the earlier example, there is one common quantity in the two sides of the equation: mass. So, if the potential and kinetic energies can mutually transform into each other, and if this transformation leads to a change in the essential characteristics of the mass in their expressions, it would be senseless to distinct the gravitational and inertial masses at all, unless one can give a physical reason on their mutual transformation into each other. So we have two options. Either to reject the distinction between the kinds of masses (and remain based on the standard physics) or to give a reasonable explanation on their possible qualitative transformation into each other. One tends to confirm the latter option, but that involves two further questions. Does such a transformation mechanism work? Do all elementary units (if, at all, there are such) of mass transform into the other form of mass? Were all of them in state m_V before the drop and will transform all of them into m_T after reaching the floor? We will return to this dilemma in Section 2.8.

We are still at the beginning and meet something unusual compared to that we learned in the school. With the cited words by L. Szilard, we face "another version" of known facts. It warns us that we should be cautious before we reduce the equation by omitting m on both sides.

2.1 Equivalence versus identity

Discussion of the relation between these two terms needs some partly philosophical approaches. In order to understand their relation, one needs to introduce another notion. Equivalence – at least in certain properties – is a symmetry relation. To associate symmetry with identity and equivalence, one cannot avoid discussing the interpretation of symmetry. Moreover, one needs to illustrate the issues on a wider scale of examples than the mass.

Ancient Greeks conceived symmetry as proportion, harmony among parts of a whole. E. Castellani (2003) interprets symmetries of a system as a whole, like equivalence relations between the elements. "The fact, that the parts are related by means of an equivalence relation (which is at the same time *the* equivalence relation between the parts, ensuring their interchangeability . . .) corresponds to the fact that the family of operations transforming the parts into each other while leaving the whole invariant satisfies the conditions for constituting a group" (p. 427). (We assume that the reader is familiar with the notion of the *group* as a well-defined algebraic structure.)

In this book, we consider the gravitational and inertial masses as parts of a mass family. Let us assume that equivalence relations in the earlier citation pertain to the two kinds of masses as well. Their equivalence should involve (does it?) a symmetry transformation (and a corresponding group) that transforms them into each other ("*the* equivalence relation which is such as to ensure the invariance of the whole when the equivalent parts are exchanged" p. 428, with the words of E. Castellani). In the following sections we will be seeking whether such symmetry exists, and if it does, what is its form.

In simple, we usually call symmetry a synthesis of constancy and change. Why?

Symmetries are closely related also to invariances (Noether, 1918; Darvas, 2018c). They are interpreted in a wide sense. The term symmetry denotes a class of phenomena, a class of properties, and a concept.

"In general terms, we speak about *symmetry* if
– under any kind of *transformation* (operation),
– at least one *property*
– of an arbitrary *object* is left invariant." (Darvas, 2002, 2007a)

Invariance under a *transformation* means the absence of change; constancy in the *properties* of a given *object*. In respect of the properties of the given "object",

- we speak about *identity*, if all of its properties are conserved (unchanged) and indistinguishable; and
- we speak about *equivalence* if certain properties of the object are conserved, but not all, and the compared states of the object (before and after the effect of the given transformation) can be distinguished.

In the case of *identity*, we speak about "the same object" before and after the effect of the transformation, which leaves it identical with itself. In other words, we say that the object saved its self-identity.

This is not so evidently unambiguous in the case of *equivalence*. In certain instances, we speak about "another object", if a transformation changed at least one of its properties (e.g. after charge conjugation, an electron will be called positron), while in other instances we speak about "the same object" (e.g. after flipping an electron's spin over, it will be called still electron) that differs from the previous one in one (or a few) of its properties, but its other properties saved their value (i.e. remained equivalent).

Let us see two examples for the ambiguity. Take in your hand a well-polished brilliant-cut diamond gem. Turn it around its main axis n times at a $2\pi/8$ angle. You find it identical with itself. You cannot distinguish the gem in its initial state and the rotated states. The indistinguishable positions are the states of the same, identical gem. If you give this gem in the hands of a jeweller, he will investigate it under a microscope and will find minor imperfections inside the gem. They are placed not symmetrically. So, what seemed identical after a rotation for you will be distinct for the jeweller. The gem will be identical with itself, but not invariant under a rotation different from 2π angle. The symmetry at an everyday scale turned out to be broken at a lower scale.

Now take in your hand a magnetised metallic cube. You can rotate it around any of its symmetry axes and will find the states (at least geometrically) identical. However, the magnetisation determines a particular (invisible) direction within the cube, and the different (rotated) states of the cube can be distinguished by their magnetic effects on other magnetisable metallic objects. The geometrically identical states of the cube will prove to be not equivalent in their (non-geometric) physical properties.

Equivalence states something in the result of a comparison between two objects. When we speak of equivalence, we compare non-identical objects or non-identical (i.e. distinguishable, two) states of an object. We can say either that two (distinct) objects are equivalent or that given properties of two objects are equivalent. In both formulations we mean, their given properties are of equal value. Equivalence presumes the existence of (at least) two objects which can be distinguished in certain properties to make us able to state the characteristic values of their other property/ies to coincide, that is, being equivalent.

In another interpretation, we can compare the states of the same (identical) object before and after a symmetry transformation, that is, when the compared things are different states of the same object observed in different times.

In a more strict sense, identity refers to the "objects" themselves, while equivalence refers to a "class of properties" of the objects.

These distinct interpretations, however, are mixed/interchanged/confused in the everyday usage of these terms. Several times we transfer this – not completely exact – everyday usage into our scientific terminology. This may lead to misunderstandings.

There is also a terminological approach to the problem. We denote identical objects by identical names. Different but equivalent objects are denoted by different names. It holds in general. However, we are not always consequent. Moreover, we must obey conventional terminology introduced in the scientific literature.

2.2 Examples of the distinction between identity and equivalence

2.2.1 Physical approach

In the following, we focus on a few specific physical examples.

In general, when two physical objects differ in their geometrical or classical mechanical properties, we consider them identical. If they differ also in other properties, we consider them different objects, between which we can ascertain the existence of an equivalence. It holds also only in general. However, physics is not consequent either.

For example, a moving point mass preserves its identity under transformations that change its velocity, its space position at different times, its linear or angular momentum, its kinetic, internal or potential energy, or all the mentioned physical properties. If a transformation changes (inverts) the spin of an electron to the opposite, the electron preserves its identity. It will continue to be called an electron. The spin, as its own angular momentum, behaves in this case similar to the classical mechanical angular momentum.

On the other hand, an object does not keep its identity when it is subject to charge conjugation, that is a non-mechanical, non-geometric transformation. If we change the negative charge of an electron for a positive one, we denote it by the term positron. Similarly, if we change the positive charge of a proton to a neutral one, we call the changed object a neutron. Furthermore, when we change the flavour of a quark, we will refer to it as a quark with another name (e.g. down, instead of up). Generally speaking, microphysical objects are classified according to various properties, mainly according to mechanical ones, like mass (hadrons, baryons, mesons, and leptons) or own angular momentum (fermions and bosons), and yet, all of them have got individual names. Unlike (e.g. in artillery) we speak of massive or less massive bullets, higher or lower speed bullets, hot or cold bullets, that means, we denote them with the common term "bullet", in our usual (particle physics) communication we do not refer to the difference between two "baryons" with qualifiers. Instead, we use their proper names. We call them, for example, a proton and a Λ^0 particle, and not "a 938.28 MeV mass, positively charged, stable baryon" and "a 1115.6 MeV mass, neutrally charged,

$2.63*10^{-10}$ s lifetime baryon". When we change the mass, velocity, temperature (internal energy), or in an extreme case even electric charge of a bullet, it preserves its identity, at least in its name; when we change the mass, energy, electric charge, and lifetime of a baryon, it loses its identity, at least it loses its name. However, in physical terms, in both cases we did the same: we executed certain transformations that changed the value of a few of their properties while left others intact.

> Looking beyond physics, chiral pairs of enantiomer molecules are considered *chemically equivalent*, although they differ in their spatial structures. A glucose molecule optically rotating the plane of the polarised light to the left (L-modification) and its chiral pair rotating it to the right (D-modification) are both glucose. However, they behave differently in biochemical reactions: D-glucose can be metabolised by living organisms, while L-glucose cannot. Can one state now that L-glucose *is identical* with D-glucose?

Returning again to physics, we assumed for a few years that we had six quarks, more precisely, we had "quark" of six different flavours. It turned out soon that each of the quarks had another property, which property could take three different values. This property was given the name colour. Then, we could say, that we have

- a *single* particle called quark, which can have six flavours and each flavoured quark can have three colours or
- *six* kinds of quarks (denoted by six different names), and each of them can appear in three colours or
- *eighteen* different quarks (denoted by 18 different names, e.g. "red charm quark"), which could be characterised by the combination of a six-valued (flavour) and a three-valued (colour) property.

One can ask, whether our standard model (SM) quarks are different particles or are they identical ones that differ in one or two of their properties? Do we have six different particles, called "up", "down", "strange", "charm", "top," and "bottom", which, moreover, appear in three different form each, called "red", "green", "blue", that means finally we encounter eighteen different particles (not mentioning their antiparticles), or do we have rather a single particle called quark, which may appear in different flavours and colours? From the aspect of their symmetries, the answer to this question is indifferent. They are obviously not identical, but are they equivalent? The answer is: in certain properties, they are, in others are not.

In mechanics, if the colour of a bullet changes (e.g. during its path along a ballistic orbit after having shot out) it will be referred to like the same bullet. If you change the flavour of your vegetable soup by salting it, it will remain a vegetable soup, although, it changed also physically (since placing the plate outside the door in winter your soup will be frozen at a lower temperature) compared to that before extra salt was added. An accelerated and rotating spaceship will remain identical with that in the previous, stable state. On the other hand, changing a property of a quark will transfer it to another named particle. Are these only merely terminological problems?

2.2.2 Mathematical approach

At the beginning of Section 2.1, we associated the notion of equivalence with invariances and symmetries. The best and conventionally accepted mathematical tools to describe symmetries are groups. From the aspect of groups and the transformations that transmit one group element into another, there makes no difference whether one denotes the state of an object after a transformation by another name or by the same. If these transformations make changes between the elements that remain within the group (do not result in elements outside the group, and meet the other axioms mathematically defined for groups) we speak about symmetries, independent of the names of the elements of the group. Our accelerated spaceship, for example, is subject of a transformation under the Lorentz group (called also $O(3,1)$ group), while a colour change between quarks is subject of a transformation under the SU(3) group. All are called symmetry transformations, by which elements belonging to the same group can be "transformed" into each other.

Are the particles that belong to the same group identical or equivalent? Are they all identical for we denote them with a common name (labelled them so that they all belong to the same group), while they must have a property according to which we can individually distinguish (i.e. identify) them? (E.g. do we consider them different, individual elements of a group, which can be distinguished from each other?) We have two ways to label them. We can denote the elements by their quality of belonging to the group, mentioning that they are equivalent in this quality, but not identical, for they differ in the value of a property, which makes them identifiable within the group. We can also denote all the elements of the group by an individual name. In physical terms, it is subject of a convention in which cases we apply this or that terminology. Anyway, we label them in this or that way, the elements of our groups *can be equivalent* in certain properties (that means, mark them to belong to the same group), but they remain distinguishable in other properties (within the group). That means, they *cannot be identical*.

We will apply this latter observation later on for the distinction of the mass of gravity and mass of inertia, then for the charges of further physical fields.

In summary, what was analogous (and what was different) with the previously learnt identities? Electrons, for example, were considered identical particles independent of the orientation that their spin took in space. Nucleons are considered identical particles with the condition that they differ only in the orientation of the isotopic spin, a property defined in an abstract gauge field. Their identity holds under the condition that they are confined within a nucleus. They lose their identity (i.e. their properties lose their equivalence) as soon as one observes them as free particles outside a nucleus.

We met two equivalent statements. In a symbolic example of two different colour cats indistinguishable at night we can say either (a) "the two cats are *different*; at best if they are locked in a dark room – where their colour, as a property, no longer plays a role – I am unable to distinguish them"; or (b) "the cats locked in the room are in fact *identical*; at best I can distinguish that one is black and the other is white in the particular event in which I go in and turn on the light". In the instance when we take nucleons into account, we can state: (a) "nucleons can be two *different* particles, called proton and neutron, which are *equivalent* in the value of all properties but their electric charge"; (b) "nucleons – confined in a nucleus – are *identical* particles, whose isotopic spin can take two orientations". Statement (a) says *equivalence of qualitatively different entities*; statement (b) says the *identity of entities*, which (in another approach) can be *characterised by quantitatively different values of a property*. Both statements hold, and it is meaningless to make difference between them from the aspect of physics.

However, there remained an open issue. Can entities be considered to be identical that are characterised by different quantities of the value of a property? My answer is no.

2.3 Equivalence does not mean identity

Let us explain why. In most cases, we can avoid ambiguity. In most of the cases, we are not concerned about all properties of the compared two objects. We speak about their *equivalence* with respect to a chosen set of properties. Equivalence means that the values assigned to the individual properties in the chosen set are equal. In this case, we either neglect or do not know their further properties. Nevertheless, equivalent objects may differ in other properties that are actually not considered. The consequence is that equivalence and identity can be applied to the same physical objects in different contexts.

In a strict sense, however, *identical objects cannot be equivalent.* Only *qualitatively different* objects can be compared to conclude a *quantitative equivalence* between them. Equivalence always presumes the existence of at least one property, in which the compared objects differ. (Compare this with the interpretation of symmetry at the beginning of Section 2.1!)

To demonstrate this, there was a good example of how the property called *colour* had been introduced in the family of quarks. A particle called Ω^- was predicted by the SU(3)-invariant model and was discovered empirically soon in the early 1960s. It was made up of three identical-flavour (*sss*) quarks. Quarks are half-integer spin fermions, whose spin can take two opposite states. According to the Pauli principle, there cannot be present two fermions in identical states in a system. The Ω^- structure contradicted this principle, for at least two *s* quarks among the three would have parallel spins. One should have assumed that the Ω^- was made up of not identical quarks, rather they differed in one quality. It was proposed that the constituents were three equivalent flavour quarks, which, however, must differ in another, yet unknown, property. This new property was later called *quark colour*. This was the way – namely *replacing an identity assumption with an equivalence, assuming another property in the background* – how quantum chromodynamics (QCD) was born. QCD is now one of the fundamental theories of the SM of physics.

The cited examples included known physical facts. They served as a conceptual introduction to expose less analysed questions in connection with the equivalence-identity problem.

2.4 The equivalence principle

2.4.1 Equivalence of the masses of gravity and inertia

We saw on the example of the necessity to introduce colours to characterise quarks that we can never be sure whether there are no further (hidden) properties in which two objects differ. In other words: equivalence does not guarantee identity.

Now, let us return to the discussion of mass. The equivalence principle is one of the main pillars of the GTR. The equivalence principle states the equivalence of the gravitational and inertial masses.

The equivalence principle (at least in its so-called Einstein's form, which lays between the weak and strong formulation, but shows more similarity to the weak) says that a test mass in a windowless box, which is affected by a mechanical force, cannot decide whether the force that he discerns originates from a gravitating object outside that box, or, at least locally, from an inertial (kinetic) acceleration source.

The equivalence principle states that the inertial mass and the gravitational mass of the test body are proportional and (as we fixed the factor of proportion to "1") are measured on the same scale. Nevertheless, they should be considered not identical properties. As we have shown, identical things cannot be equivalent. Equivalence is a quantitative relation between qualitatively different (non-identical) entities. Only different things can be compared and proven to be equivalent. One needs to have two *different qualities* to claim they are of *equivalent quantities*.

Many physics textbooks conclude from the equivalence principle the identity of the gravitational mass and the inertial mass. We will see in Section 2.5 that Einstein originally assumed the same, and he changed his mind a few years later only. (That change did not destroy the validity of the GTR.) We will argue for they are not identical. Nevertheless, they are different properties of the matter, not only because we denote them by different names. The essence to establish the principle, and to demonstrate experimentally the equivalence, was that we had gained experience about the two mass properties from different observations. The goal of the experiments to demonstrate their equivalence was to construct measurements, where both properties are present. We must emphasize both. We speak about "both" in the case of qualitatively different entities. Another remark: the measurements were performed in rest. The gravitational mass and the inertial mass are qualitatively different entities, which proved to be equivalent in the measure of their effects. They are equivalent in their value (the measure of mass), but they are not identical properties.

We will list a few consequences of the non-identity of the equivalent masses. For school-level calculation purposes, it was enough to denote both masses with the identical character "m". This can be justified, for their values are equal (at least, while the relative velocity of the compared masses is not too large). However, this equality cannot satisfy our curiosity about certain theoretical (not just mathematical), philosophical, and conceptual consequences.

At this stage, let us say that we consider the two kinds of masses as siblings.

2.5 Transformation properties of gravitational and inertial masses

The distinction between the two kinds of masses, $m_{gravity}=m_V$ and $m_{inertia}=m_T$ is not new in the history of physics. Their difference was known since Newton. However, as shown in Section 2, due to their quantitative equivalence (at least at rest) it was not necessary to distinguish them qualitatively in the equations of physics. So they were identified for long. This was pragmatically correct, but theoretically not justified. The problem became acute when technological development made possible to observe experimentally high energy interactions.

There are different approaches to handle the problem of the two kinds of masses. Let us mention first the original approach by Einstein. In the 1910s, he considered that according to GTR, there were only free falling bodies and inertial forces in the gravitational field, therefore the notion of the gravitational mass could be cancelled from the vocabulary of physics, and anyway, if they both existed, they were equivalent. The reason of this approach was that the left side of the gravitational equations includes only geometric properties of the curved space–time, so there is no place for masses (field sources) on the right side as well. Later he changed his mind and noted that the individual elements of the stress–energy tensor were associated with energy and momentum (or their densities), that means, with m_V and m_T, respectively; and the gravitational constant on the right side compensated the mass dimensions.

From our point of view, the most important approaches are those that deal with the *transformation* of the two kinds of masses. We saw in Section 2 that their equivalence should be associated with transformation properties. There is an opinion that the masses do not transform with the velocity (Hraskó, 2001). The change of their inertia is due to the Lorentz transformation of their velocity, and it cannot be taken into consideration twice. Moreover, the inertia of a moving body transforms in a different proportion in the direction of its velocity (longitudinal inertial mass) and perpendicular to it (transversal inertial mass), what is an argument for leaving the Lorentz transformation solely to the velocity, and against the transformation of the mass itself, that is, to calculate always with the rest mass. Nevertheless, this argument does not eliminate the qualitative difference between the masses m_V and m_T. We will reflect the separation of the roles of the masses and velocities in Sections 4.4 and 12.3.

One finds in many textbooks that in the equation $E = mc^2$ letter “m” denotes the *inertial mass*. This holds only for calculation purposes since any amount of gravitational mass is equivalent with the same amount of inertial mass (let us add: in rest), so quantitatively it can be replaced by that. In his first mention of this equation in 1905, Einstein meant some internal energy (not identical with the thermodynamical meaning of this quantity) of a “moving body”. In his next, 1905 September paper, Einstein mentioned the inertial mass (appearing in the right side). When he formulated the GTR 10 years later, he corrected himself and identified the mass in this equation with the gravitational mass. Even in 1921, Einstein himself wrote (p. 783): “If an amount of energy E be given to a body, the *internal* mass of the body increases by an amount E/c^2.” (Emphasis by me – G.D.) This indicates that – at least in the first period – even the father of the idea was not certain about the nature of this statement. Von Laue (1911 and 1955) emphasised the restricted validity of Einstein’s $E=mc^2$ too. He realised at an early stage that it holds only in static systems and in which pressure vanishes. In any other case neither the energy (left side), nor the mass (at the right side) cannot be handled as homogenous, indivisible properties.

A recent another approach to make a distinction between forms of masses is discussed by Calmet and Kuntz (2017): According to them a few observed phenomena “suggest that there is a new form of matter that does not shine in the electromagnetic spectrum. Dark matter is not accounted for by either general relativity or the standard model of particle physics. While a large fraction of the high energy community is convinced that dark matter should be described by yet undiscovered new particles, it remains an open question whether this phenomenon requires a modification of the standard model or general relativity. Here we want to raise a slightly different question namely whether the distinction between modified gravity or new particles is always clear.” They showed that this is not always the case. My comment here is to refer to the mass of the dion, mentioned in Section 11.3.14.

E.P.J. de Haas referred (de Haas, 2004a, 2005) to an early assumption by G. Mie (1912a, 1912b, 1913) concerning the transformation of the masses. The papers published by Mie contributed to the elaboration of the gravitational theory under preparation at that time. According to Mie, the inertial mass transforms as $m_T=\kappa m_0$, and the gravitational mass transforms as $m_V= (1/\kappa)m_0$ where m_0 denotes the rest mass, and κ represents its Lorentz transformation. Based on these transformations, Mie formulated a version of the weak equivalence principle. (It was never accepted by Einstein.) As de Haas explicated in another paper (de Haas, 2004b), Mie’s idea complied with a later hypothesis by L. de Broglie (1923, 1925) on waves connected to particles with material mass. de Broglie's theory assigned wave frequencies to the moving massive particles that he called inertial clock frequencies (ν_T) and inner-clock frequencies (ν_V). They transform as $\nu_T = \kappa\nu_0$ and $\nu_V = (1/\kappa)\nu_0$, respectively. Although this assumption included a contradiction, de Broglie solved it by his so-called "Harmony of the Phases". In that theory, de Broglie first stated that the gravitational mass and the inertial mass of the same particle may possess particle- and wave properties, respectively. (This holds even in contemporary theories.) de Haas showed that accepting the transformations assumed by Mie and de Broglie lead to the observations that, seen from a rest frame, "the equivalence of the masses is not a Lorentz-invariant condition and cannot be transformed into a fundamental axiom or law of nature"; and the same observer may conclude that "the equivalence of the phases is a Lorentz-invariant condition and that this equivalence can be seen as a fundamental law of nature". In short, he concluded that the equivalence principle of masses should be replaced by the equivalence principle of wave phases. de Haas mentioned also that answering the questions raised by the mentioned controversies is left for quantum gravity.

We mention another, widely circulated approach, found in a few Internet sources (by S. Fedosin, Perm) that argues repeatedly for a fixed 4/3 proportion between the masses m_T and m_V. This approach can hardly be justified, at least according to the author of this book.

Accepting the usual form of the GTR (e.g. G. 't Hooft, 2002), maintaining the validity of the weak equivalence principle for masses, requires to reject the assumed Mie–de Broglie transformation rules. One can check it in an easy thought experiment. Let us imagine two, distant, different mass celestial bodies revolving around their mutual mass centre at high velocities, in respect both to each other and to the Earth. Compare how do they observe the motion of each other from their reference frames, and how do we observe their motion from the Earth. (In fact, we can observe the period of the smaller one from the Earth.) Assuming the Mie–de Broglie transformation, we get in contradiction with the values expected by the special theory of relativity (STR) and the Newton's gravitational law. Our observations will coincide with those laws assuming the following transformations: $m_V = m_0$ and $m_T = \kappa m_0$, where m_0 is the measure of the rest mass.

According to assumptions in the theory applied in this book, the gravitational mass does not change with velocity boost, while the inertial mass Lorentz-transforms. Let us remember also a consideration that in the same theory (Darvas, 2011) interacting masses must be always in opposite sibling states: m_V can interact with m_T and vice versa. (We will discuss this mechanism in Section 10.1.) As we assume, they can be transformed into each other (see Section 11.2), since they are subject of an invariance under the transformations of a symmetry group (to be described in Section 5), resulted in a conservation [analytically Sections 7 and 8 and (Darvas, 2011), algebraically Section 5.2.5 and (Darvas, 2015c, 2018b)]. Their roles are changed permanently, like fermions do with their boson exchange according to the interaction mechanism to be described in Section 10.1 [see also (Darvas, 2012a) specification to the gravity in Sections 2.6 and 11.2, while a Lagrangian description is given in (Darvas, 2012b, Appendix)].

In short, the transformation rules of the gravitational and the inertial masses differ from each other. The gravitational mass does not change with the velocity, while the inertial mass increases with the Lorentz transformation during a velocity boost. Therefore, the two forms of mass cannot be distinguished in rest and at low velocities, although they significantly differ near to the speed of light.

More than a hundred years after the works of Eötvös, von Laue, Mie, Hilbert, Einstein and others, one can no more be in doubt the different transformation of the two kinds of mass. One may have concerns whether which of the revealed transformation rules is the proper one. However, one must admit the fact of the disparate transformations that the mass siblings obey. The distinct transformation rules form a strong argument for the qualitative difference between the gravitational and the inertial masses.

2.6 The role of masses in the stress–energy tensor of GTR

The role of the mass is replaced by the stress–energy tensor in GTR. Before applying the mass siblings for the stress–energy tensor, we should make a few further preliminary remarks. The described assumptions in this book (including, among others, the distinction between the gravitational and inertial masses) are relevant among strongly relativistic conditions. (We speak about strongly relativistic conditions at high kinetic energies with velocities approaching the speed of light.) The derivation of the GTR included a few approximations assuming relatively weak gravitational field and not too large velocities. Similar approximations were assumed in the solutions of the Einstein equations. And yet, seemingly, the theory works well among wide limits. One can apply the distinction between the mass siblings by accepting that the Einstein equations hold (at least approximately) in the presence of strong gravitation and masses moving at high speed. Nevertheless, in the latter instance, we should consider certain extensions of the theory to strong gravitation and the presence of a velocity-dependent field effective at high energies.

The components of the stress–energy tensor can be clustered in four parts.

stress density

$$\left[\begin{array}{ccc:c} T_{11} & T_{12} & T_{13} & T_{14} \\ T_{21} & T_{22} & T_{23} & T_{24} \\ T_{31} & T_{32} & T_{33} & T_{34} \\ \hdashline T_{41} & T_{42} & T_{43} & T_{44} \end{array}\right] \begin{array}{l} \left.\begin{array}{l} \\ \\ \\ \end{array}\right\} \text{momentum density} \\ \\ \end{array} \tag{2.1}$$

energy flux density energy density

The stress–energy tensor is symmetric. This means, the respective elements of the momentum density and the energy flux density are equal: $T_{i4} = \delta^{ik} T_{4k}$.

The assumption of the difference between the gravitational and inertial masses, apparently, distorts this symmetry. There appear inertial masses (m_T) in the stress- and the momentum density, while there appear gravitational masses (m_V) in the energy flux density and the energy density. (See also a few additional remarks on this categorisation in Section 11.2.1.) The former transform with velocity, while the latter do not.

We intend to keep the symmetry of the stress–energy tensor. In this order, we should demand that under a velocity boost, the tensor transformed so that along with the invariance under the Lorentz transformation it transformed also in an invariant way under another transformation, which could transform the two kinds of masses into each other. This latter transformation will be derived in Section 11.2.1 [and has been presented in several former publications by the author (Darvas, 2011; 2012a; 2012b; 2012c; 2017c)]. This combined transformation – invariance under a symmetry group to be introduced – guarantees to keep the symmetry of the tensor out of rest. However, this transformation is unusual in the traditional, standard physics.

2.7 Conservation of mass

Let us investigate the conservation of mass. It has been assumed an apparently unproblematic question, without any open problem in connection with it. Nevertheless, the picture is not so simple.

How did we conclude the mass conservation? In classical mechanics, we had empirical evidence for the conservation of energy. We had also empirical evidence for the conservation of mass (in general). Then three new issues entered the scene.

(i) Proportionality between the quantity of energy and the measured quantity of mass was established.
(ii) We have got also proportionality between the measured quantities of the gravitational mass and the inertial mass.
(iii) Finally, we have a principle of equivalence.

Thus, we concluded from the conservation of energy the conservation of mass, and through the proportionality between the two kinds of masses and applying the equivalence principle, we extended the conservation to all kinds of masses.

Let us reconsider this logic. We must mention in advance the problem that in (*i*) the energy was not the potential or the kinetic energy, rather the internal energy of a system (Einstein, 1905a, b; Hraskó, 2003, see also in Section 11.2.1); the mass in the equation $E=mc^2$ was identical with the gravitational mass (see also in Section 2.5). The conservation of the energy (like other mechanical quantities) was concluded from the integration of the equations of motion. In modern treatment, we can obtain it by the variation of the Lagrangian $L(x, \dot{x}, t)$ for the geometric invariances (applying Noether's first theorem). The conserved energy, what we get, is proportional to the *mass* of the investigated system (or the whole universe). To *which* mass? To the *gravitational mass*. Where do we deduce from, that the full mass is conserved? We conclude it from the principle of equivalence. What does the principle of equivalence say us? It says, that

(a) The effects of the two types of mass are indistinguishable.
(b) Moreover, we knew earlier that the mass of a given object can behave both like gravitational mass and inertial mass.
(c) The measured quantities of these two masses are equal.

Sorry to say, these together are logically inadequate to conclude the conservation of the full mass.

If we assume that the inertial and gravitational masses are two qualitatively different properties of matter (at least, based on the listed clue), we have no reason to make any statement on the conservation of the inertial mass. The quantities of the two masses are equal, but they are supposed to be not identical.[2.1] This means that we concluded the conservation of the gravitational mass (from the conservation of the energy), and we have good reason to state that this conserved amount gravitational mass is in its quantity equivalent to a certain amount of inertial mass. No more. It does not follow from this conclusion that there are no other quantities of inertial mass in our universe that are not without doubt conserved. I do not state that certainly there are such non-conserved (inertial) masses. I state only that all

2.1 Einstein (1921, p. 783) wrote: " . . . the gravitational and inertial masses of a body are numerically equal to each other. This numerical equality suggests identity in character. Can gravitation and inertia be identical?" Einstein felt convinced of the identity of inertial and gravitational mass on the basis of the success of the GTR, for "the same property which is regarded as *inertia* from the point of view of a system not taking part in the rotation can be interpreted as *gravitation* when considered with respect to a system that shares the rotation." This is nothing else than his comprehension of *covariance* in GTR. Einstein considered the acceptance of the identity as a caesura between the Newtonian theory and GTR. We do not share fully this interpretation by Einstein: "numerically equal" does not mean identity, only equivalence in their measure and effect, as we explained. Even the question about their identity marks that, in fact, he meant equivalence.

De Haas, E.P.J. (2004a) identifies the behaviour of the *gravitational mass* interpreted by Mie (1912a) with the clock-like frequency of de Broglie (1923) and Mie's *inertial mass* with de Broglie's wave-like frequency. De Haas concludes that de Broglie's "harmony of the phases" should be interpreted as an equivalence principle for quantum gravity, thus giving base for an interpretation of the *particle–wave duality*. What most important from our aspect is, according to De Haas, E.P.J. (2004a), the Mie–de Broglie interpretation "suggests a correction of Hamilton's variational principle in the quantum domain. The equivalence of the masses can be seen as a classical 'limit' of the quantum equivalence of the phases." He establishes the statement that gravity seems to be a particle-like aspect of elementary particles, and inertia a wave-like aspect, in the framework of the *weak equivalence principle* in the Mie–Einstein meaning.

the earlier conclusions did not provide evidence for it. It has not been proven.[2.2] This is a very strong statement that deserves further attention, and another argument for the distinction between the members of the mass siblings.

If we want to find evidence for the conservation of the full mass (both the gravitational and inertial, similar to electromagnetics and the conservation of the electric charge), we should turn to the four-potential of the gravitational field and the energy–momentum tensor introduced in general relativity. In this course, it is irrespective that the mass is a quite different property "charge" of the field equations, quite different bosons mediate their interactions, and quite different Lagrangians govern their states and interactions than the electric charges in the electromagnetic field. The common feature between them is the role of a central ($\sim 1/r$) scalar potential plus a kinetic part and that we should expect some gauge invariance resulted from the four-potential. This latter did not follow from classical mechanics. Nevertheless, the earlier conclusion on the restriction of conservation to the gravitational mass did not go beyond classical mechanics.

When we concluded the conservation of the mass solely from the gravitational potential, we ignored any possible contribution by the kinetic part of the Hamiltonian (while the full Hamiltonian was generated by the energy–momentum tensor in the general theory of gravitation). Similarly, like we derived the conservation of the electric charge – in classical electrodynamics – from the Maxwell equations alone, we derived an invariance solely from a transformation in the Coulomb field. Thus – in classical electrodynamics – we did not couple it with a transformation in the gauge field (which latter was generated by the rest of the electromagnetic field tensor, and what we will see in Section 3.5). This latter "imperfection" has been corrected by the coupled gauge transformation[2.3] in quantum electrodynamics (QED). Similar "correction" is to be done in case of the conservation of mass (cf., Section 11.2).

2.8 Some preliminary consequences of the distinction between masses of gravity and inertia

There arises the question: if our concepts on the conservation of mass were not perfect, how could we derive proper conservation laws?

Furthermore, is this the only consequence of the distinction? It is only one side of the problem to consider analogies to the method we have learned in "classical" QED. We have to overview all of our equations where mass appears and replace the

2.2 For example, let us imagine a dance school. Boys and girls attend this school. The music starts and all the boys invite girls to dance. The observer registers that all boys have found a partner. Then we read the record. Can we state that there were no more girls in the school?

2.3 In a proper gauge theory, symmetry transformations leave the total Hamiltonian invariant, and not the kinetic and the potential components of the energy separately.

general mass term m either with the gravitational mass m_V or with the inertial mass m_T. This change $f(m) \rightarrow f(m_V, m_T)$ will not influence the result of any previously performed calculations since the numerical values of m_V and m_T are equal. However, this change can make alterations in the formulation of certain laws, and not only those on conservation.

To make the program more transparent, we give a few examples for the replacement.

There is a common school task to determine the ballistic orbit of a bullet. This bullet moves as a result of the initial momentum it got when shot out from the gun, and the gravitational force of the Earth. Its inertial mass appears in its momentum, while its gravitational mass in the effect originating from the Earth.

As a next example, let us take a look at a more general situation. We have an equation (without giving them concrete physical meaning): $F(x,t) = m \cdot g(x,t) + m \cdot h(x,t)$, where F, g, and h are general, hypothetic functions. We learned in the school that we can make the following calculations: $F(x,t) = m \cdot [g(x,t) + h(x,t)]$ and divide both sides by m and get:

$$\frac{F(x,t)}{m} = g(x,t) + h(x,t)$$

The numerical value of the two sides of the first form of the equation does not change, if we replace m with m_V and m_T, respectively. The situation becomes interesting if there appear both masses in the equation: $F(x,t) = m_V \cdot g(x,t) + m_T \cdot h(x,t)$ like in the case of the shot bullet. In this instance, we will no more be allowed to introduce the two steps in the calculation as shown earlier.

Third, let us turn to a more advanced problem, namely to derive a kind of conservation. For this reason, we must calculate the variation of the action of the investigated system's Lagrangian (Darvas, 2018c). In general, there will appear a potential part and a kinetic part in the Lagrangian (and possibly others, to describe interactions). In the simplest case, the potential part will include gravitational mass and the kinetic part inertial mass, but in less simple cases, the situation can be even more complicated with additional components. In this example, the final result can be influenced by the difference caused by the presence of the two kinds of masses.

This last consideration may bring new issues if we repeat the derivation of our long-established laws of physics. This is surprising in itself that it has been missing, is it not? Perhaps, this can lead us to formulate the real conservation law for masses. However, that will be part of a new approach to physics that introduces the two kinds of masses as distinct physical quantities.

2.8.1 Where can one meet separated inertial and gravitational masses?

As we saw in the third example, the inertial masses appear in the kinetic part of the Lagrangian of a physical system and the gravitational masses in the potential part of the same Lagrangian. They appear similarly in the Hamiltonian of a physical system: the inertial masses in the kinetic part and the gravitational masses in the potential part. The kinetic part of a Hamiltonian is called also vector part, while the potential part – scalar. The scalar (inertial) masses appear in the kinetic part as coefficients of vector components, namely in mechanical systems associated with the velocity vector, while also the scalar (gravitational) masses appear in the potential part as coefficients of scalar quantities. These mass siblings together form the source of the gravitational interaction field and are called field sources (cf., Section 4.1).

They appear similarly in the four-momentum. In the general theory of gravitation, the stress–energy tensor plays the role of the mass. The stress–energy tensor is often called momentum–energy tensor. In (+, +, +, −) signature, the first three rows of the momentum–energy tensor [cf., Formula (2.1)] are associated with the inertial mass, while the fourth row with the gravitational mass. We will show in Section 11.2.2, that one needs an additional symmetry transformation to restore the so violated symmetry of the tensor.

3 ELECTRIC CHARGE

3.1 Distinction between electric charges

The classical papers by Dirac (1928, 1929) that based quantum electrodynamics (QED) were formulated by him as "the theory of the electron". He used this phrase even in his later papers (1951a; 1951b; 1962). Several leading physicists of the time followed him in the years 1929–1932.

This list included, among others, Fermi and Breit. Fermi opposed Dirac's approach in his own derivations of QED since the beginning. And yet, he also started from an interaction between Coulomb potentials and considered the effect of a vector potential in a perturbation (Fermi, 1931, 1932). So did Breit (1929, 1932). All they started from a picture in which they supposed that initially, the (static, scalar) Coulomb potentials of the electric charges interact. (Coulomb charges are considered to be in bound states.) Then the Coulomb charges bring each other in motion and that propagates the vector potentials and interaction of those too. (Lorentz charges appearing in the vector potentials are considered to be in free states.) For example, Dirac applied radiation gauge, and in zero approximation the unperturbed Hamiltonian contained the Coulomb interaction only, the next parts of the perturbation introduced the effects of the vector potential. Heisenberg (1931) supported the same model of perturbation.

C. Møller (1931) proposed an opposite way. He considered a kinetic interaction in the unperturbed Hamiltonian and took into account the Coulomb interaction in the perturbation. His approach was not widely accepted, for it included an *asymmetry in the role of the interacting charges.*

In 1932, Bethe proposed Fermi to describe the interaction between two charged particles assuming that they both were initially in free-particle states (and to apply Lorentz gauge) so that they modified the model proposed by Møller. They introduced a(n artificial) *symmetrisation* in the roles of the interacting particles (Bethe, Fermi, 1932). In contrast to the derivation by Breit, the Coulomb energy in Møller's theory is a part of the perturbation and enters additionally on the side of the interaction with the (electromagnetic) tension field that was considered initially in the unperturbed approximation (Bethe, Fermi, 1932, p. 20) (i.e. in zero approximation the two electrons behave as independent from each other). They build the retarded potentials of the electrons, where the field of the first electron influences the second electron as a perturbing effect. Then they calculate the matrix elements of the interaction energy of the exchange between two states of the first electron (n_1, n'_1) and two other states of the concerned second electron (n_2, n'_2) resulted in the interaction, while the sum of their energies is conserved. The calculated matrix element

https://doi.org/10.1515/9783110713183-003

describing the transition of the second electron from state n_2 to state n'_2 by Møller is as follows:

$$V_{n_1 n_2}^{n'_1 n'_2} = e_1 e_2 e^{\frac{2\pi i}{h}(E'_1 + E'_2 - E_1 - E_2)t}.$$

$$\cdot \int u_2^{\prime *}(\vec{r}_2) u_1^{\prime *}(\vec{r}_1) \begin{bmatrix} \frac{1}{|\vec{r}_2 - \vec{r}_1|} - \\ - \frac{2\pi^2}{h^2 c^2}(E_1 - E'_1)^2 |\vec{r}_2 - \vec{r}_1| - \\ - \frac{(\gamma_1 \gamma_2)}{|\vec{r}_2 - \vec{r}_1|} \end{bmatrix} u_2(\vec{r}_2) u_1(\vec{r}_1) d\tau_1 d\tau_2$$

The first expression (upper row) $\frac{1}{|\vec{r}_2 - \vec{r}_1|}$ in the square bracket [ ] denotes the corresponding Coulomb potential, the second originates in the retardation of the scalar potential, and the third is the effect of the (unretarded in first approximation) vector potential. Bethe and Fermi found that the first and the third expressions were symmetric for the two electrons but the second was asymmetric. It depended on the energy change in the states of one of the electrons only. They ascertained that this asymmetry was a result of incompleteness of the method applied by Møller. Therefore, regarding the requirement that the full energy in the starting and final states should be equal, they artificially made that expression symmetric and replaced $(E_1 - E'_1)^2$ by $-(E_1 - E'_1)(E_2 - E'_2)$. Nothing justified this artificial step but the symmetry principle that prevailed in science since Herodotos to Kepler (Darvas, 2007a) and often even later in physics.

Without continuing their clue, in my opinion, this artificial symmetrisation is nice, but it is misleading and physically not well founded. Bethe and Fermi forced an assumption upon Møller, which did not constitute a part of Møller's theory, namely that the two interacting particles must play equal roles. Starting from this point of the derivation by Bethe and Fermi, we must separate their clue from the clue of Møller. Bethe and Fermi insisted on the symmetric role of two interacting particles, while Møller foresaw (or intuitively felt) that two interacting particles could be in two different states, although he could not clearly identify the essence of the distinction between the roles of the interacting agents. In my opinion, we do not need to demand that the individual interaction potentials be symmetric in respect of the two interacting agents. (I assume this claim is following those put down by Møller, although we are unable to check whether this was really his conscious intention.) We need to demand only that in an opposite situation – that means when the second particle plays the active role and the former is the passive (this asymmetry can be exemplified by the emission and absorption of a photon), or mirror scattering – similar (symmetric) potentials be valid and their numeric values coincide with those in the first situation. We will return to the possible mechanism of the interaction between the two charges in Section 11.1 and add there some comments to the mechanism mentioned in the second and third paragraphs of this section.

However, the main conclusion of Bethe and Fermi concentrated not on the unperturbed kinetic model, rather on proving that the model proposed by Møller is an equivalent description with the former Dirac–Fermi–Breit-type description of the interaction between two electrons. In Møller's picture, the interaction of two approaching particles starts from infinity, where the Coulomb fields affecting each other are weak, and thus, can be neglected. In 1932, Bethe and Fermi used perturbation theory too, where the scalar potential appears in the next approximations when the interacting particles get closer to each other in a process of scattering. The symmetrisation used by them – and which was introduced to the equation artificially – fitted better in the picture about the mechanism of interactions prevailing in the early 1930s, although it was shown later that their calculations lead to deviations from the experimental data (Araki and Huzinaga, 1951; Salpeter and Bethe, 1951).

In short, we had two models in classical QED. One, in which the interaction starts between two bound states of particles, and another, in which the interaction starts between two free state particles. In the first model, the effect of the vector potential enters only in the perturbation; in the second model, the role of the scalar potential is left for the perturbation. The main conclusion of the Bethe–Fermi paper (1932) was that the two descriptions are equivalent. (For more details about the birth of these different approaches to QED, see Schweber (2002) and Dyson (2005).) Both pictures are approximations, and both lead to relatively good results in accordance with the experience.

All pioneers of "classical" QED insisted on the paradigm that the roles of two interacting particles should be symmetrical, but C. Møller. He calculated scattering matrix elements between two interacting electrons and those were asymmetric in respect of the two interacting particles. The voluntary involvement by Bethe and Fermi (1932) in the equation has been justified by the symmetry paradigm only. Nevertheless, providing that two interacting electrons form a system (at least during their interaction), this symmetry, that is assumption of the identical state of the two interacting electrons within that system, contradicted the Pauli exclusion principle. Symmetry principle has been prevailing scientific thinking for over 24 centuries up to now. Kepler was the first who gave up his strong belief in the symmetry principle for the arguments of empirical data. Belief in symmetry is often stronger than any other reason, even now. According to our present knowledge, the Pauli principle proved to be stronger than the symmetry principle. And yet, due to the later high respect of both authors, the legitimacy of this artificial symmetrisation was not questioned for seven decades. Its history appeared again in the physical literature only in the early 2000s, just in short after the theory described in this book started to be developed. Discussion of the Møller scattering is a spreading subject in the physical literature in different chapters of field theories, since that period. To keep the validity of the Pauli principle but save also our belief in symmetry, we will apply another form of it, called hypersymmetry that acts at a still hidden level.

Nevertheless, the model proposed by the Møller scattering matrix suggests a further conclusion: a distinction between two states of the electric charge. We will show an example of the application of this distinction in Section 3.5.

3.2 Sources of the electromagnetic field

First, note the covariance of the electromagnetic field, namely, what could look like an electric field for one observer could be a superposition of an electric and a magnetic field for the other.[3.1] In an idealised case, if the velocity of the moving charges for the latter observer's reference frame is relativistically large, the effects of the electric field can be neglected compared to that of the magnetic field, since in that instance the effects of the electric field are magnitudes weaker. The electric field's effect is formally analogous to the gravitational effect (derived from the potential part of the Hamiltonian, cf., the Coulomb force), while the effect originating from the magnetic field shows similarity to the inertial effect in mechanics (derived from a kinetic part of the Hamiltonian; cf., the Lorentz force).

A q test charge, which is affected by an electromagnetic force, at least locally, in a small environment, will not be able to make distinction between the sources of the force that he discerns (the causes of its acceleration), whether those are caused by the electric force of another static charge, or by a magnetic force that originates from a high-speed moving charge, a current, or a magnetic field of a different source. This is very similar to the (weak) equivalence principle formulated earlier for masses.

To see it in more detail, let us take a unit test charge and fix it to the origin of our reference frame, where it is at rest. At a given moment, our test charge experiences the action of an electromagnetic force. Where can this action originate from? An acceptable simplified answer is that it comes from another charge. For simplicity, let us restrict our investigation to the case where the experienced effect is caused by a single, point-like ("agent") charge. In this case, the force can be the resultant of two types of forces: a Coulomb force, on the one hand, and (provided that the agent charge moves) a Lorentz force, on the other. Generally, the given resultant force can be composed of various pairs of Coulomb and Lorentz forces, according to the rule of vector addition. What does the magnitude and direction of

---

3.1 This demonstrative formulation refers to 't Hooft (2002). Nevertheless, the conceptual interpretation of covariance evoked much debates in twentieth-century physics. Wilczek (1998) noticed that "when physicists refer to general covariance, they usually mean the form-invariance of physical laws under coordinate transformations following the usual laws of tensor calculus, including the transformation of a given, preferred metric tensor. ... From a purely mathematical point of view one might consider doing without the metric tensor; in that case general covariance becomes essentially the same concept as topological invariance." He also reminds that in the understanding that Abelian gauge invariance and in its non-Abelian generalization – as a symmetry under transformations of quantum-mechanical wave functions – "the space-time aspect is lost. The gauge transformations act only on internal variables." In our following treatment, space–time co-ordinates take a role of internal variables. Topological aspect is considered in the sense of the generalised variables in the second Noether's theorem (cf., Darvas, 2018c). For the specification of these internal variables in this generalised sense, see the detailed presentation of the proposed theory in Section 7.3.

these forces depend on? They depend partly on the place of the agent charge at the moment when the action originated. Second, since the action travels at a finite velocity (in vacuum at the speed of light) they depend, functioning on the agent's distance from the origin of the reference frame, on the time, how much earlier the agent charge emitted the action before detection by the test charge. Third, it depends on the magnitude of the agent charge; and fourth, on the velocity of the agent charge in proportion to the test charge. Finally, it depends on the velocity of travel of the "action" in the given medium matter (as determined by the dielectric constant and magnetic permeability of the medium). Taking into account those parameters, the magnitude and direction of the forces on the test charge depend on the following: three space co-ordinates, one time co-ordinate, three velocity co-ordinates, one value for the speed of light in the given medium, and one value for the electric charge. These make altogether nine parameters; representing geometric ones, on the one hand, and characteristics of the electromagnetic field, on the other. There are interrelations among the nine parameters (via the Coulomb law, Lorentz force, etc.), but they do not unambiguously determine all nine parameters. The number of parameters is larger than the number of equations relating to them. If one fixes a few of those parameters, one can find others (free parameters) among them that can be independently varied. This "independent" variation is, however, not unlimited. For example, although one can produce an action by a given source (agent) charge, and one can instead evoke the same action on the test charge by changing (varying) the magnitude of that source charge, but one must do it for the source charge being at another place, and emitting at another time, or having another velocity. Also, if one varies the velocity of the source charge, then one should change the magnitude of the charge, as well as the place and the time of the emission of the action. All these variations should be done while taking into account the known relations (equations) among the parameters. The question is then: What is the *locus* of the set of parameters ("points") in the spacetime-electromagnetic field conglomerate that could evoke the experienced action? There exist infinitely many such "points" (sets of parameters), but the relations among them do not allow all arbitrary parameter combinations, those relations limit their set. The force experienced by the test charge is *invariant* under some selection from this *locus* of the sets of parameters ("points"); in other words, "where" the experienced action originates. But the test charge cannot discern which element of the set of the allowed events caused the given action. Hence, the given action is *invariant* under a choice from that set. Many causes may evoke the same effect. The origin behind all of them is the combination of a Coulomb force and a Lorentz force. Both originate from the same "point" in the spacetime-electromagnetic field conglomerate, although one can question the identity of the sources (agent charges) of the two forces (Darvas, 2018c). Of course, they are equivalent (in their quantity) according to the covariance principle, but, as we know, equivalence does not provide automatically guarantee for identity.

Depending on the chosen system of reference, under the covariance principle, the same charge can be in rest or moving at high velocity. Therefore, the same object can behave once as the source of a Coulomb force, and in another frame of reference as a source of a (kinetic, Lorentz) magnetic force. We can consider this so that the same agent can be either a source of a Coulomb force or the source of a magnetic force, and *these are two properties of the given charge*. In other words, *they have two states of the same property* (i.e. of the charge) of the object.

3.3 Equivalence principle for electric charges

We can formulate an *equivalence principle for electric charges*, which means that charges in the two states are equivalent in their charge value. At the same time, they are not considered identical physical properties (cf., Section 2.3). They are two different qualities, and the same object can occupy any of the two states, namely a potential, static, or Coulomb charge, and a kinetic, current-like (Lorentz) charge. Since the electric charge is the field-charge of the electromagnetic field, its two states compose together the field-charges of the electromagnetic field. Note that both negative and positive electric charges can occupy both states, respectively. The charge components together form a four-current. In that current, the Coulomb charge is a coefficient of a scalar property, while the current-like charges are coefficients of a three-component vector.

Looking for analogies, note, the Coulomb charge (as electric field-charge) and the gravitational mass (as gravitational field-charge) are associated with the scalar potentials of the given fields. The charges composing electric current can be associated with the vector potentials of the electromagnetic (gauge) field like we assumed the inertial mass in a mechanical kinetic field. The latter do not serve as sources of the (scalar) electric and gravitational fields, respectively.

Accordingly, in analogy with what we learned about mass(es), let us denote the charges appearing in the scalar potentials (V) by q_V, and the charges associated with the kinetic (vector) potentials (T) by q_T.

3.4 Transformation of the two types of electromagnetic charges

As we saw in Section 2.5 on mass, the gravitational and the inertial masses are subjects of different rules observed at increasing velocity. We have no similar experience concerning the Coulomb and Lorentz charges. Nevertheless, this observation holds when one observes the charges as field sources by themselves. The (magnetic) effect of a Lorentz charge moving at high velocity in a given reference frame will be stronger, but the cause of this increase is not an increase in its charge, instead, it is caused by the velocity. And yet, there is a difference in their transformation properties.

The field-charges of the electromagnetic field appear in the four-currents as densities. *Current densities* follow similar transformation rules like the gravitational and the inertial masses do, respectively. The constituent charge densities transform according to Lorentz "dilation" (increase) in density (contraction in diameter), in the instance of the Lorentz charges, while the charge density of the Coulomb charge, which is interpreted in rest in its reference frame, keeps its value.

3.5 Some preliminary consequences of the distinction between field-charges of the electromagnetic field

We extend our separation programme (formulated first for the charges of the gravitational field) to the charges of the electromagnetic field. This means, we replace the charges appearing in our physical equations by two different charges: a Coulomb-type one (q_V), and a kinetic-type one (q_T). We similarly describe the electromagnetic field, as we did in the case of classical electromagnetic currents, only now we extend the distinction with the introduction of the two kinds of electromagnetic charges. Here, we give a preliminary example for the implementation of this replacement.

In classical electrodynamics, the A_μ four-potentials of the electromagnetic field were invariant under Lorentz transformation and the j_ν four-current components transformed like a four-vector. Now, we assume that the sources of the Coulomb force (q_V) are different type charges than moving charges as sources of currents (q_T type charges where the index T refers to their kinetic character). The same charges play both roles (cf., covariance), we assume only, that in the two situations they behave as two states of the same physical property (i.e. field-charge). Provided that the fourth component (in + + + – signature) of the j_ν current density, namely j_4=$ic\rho$, contains a different kind of charge density (ρ_V), compared to those moving (current-like, kinetic) charges (ρ_T) in j_i (i = 1, 2, 3), the **j** four-current would lose its invariance under Lorentz transformation. Since we do not intend to lose the Lorentz invariance of long-standing, this is a first but strong symptom that calls to find some compensation solution (i.e. to restore invariance under transformation).

We can demonstrate this through the transformation of the Lorentz force: $F^\mu = (1/c)F^{\mu\nu}j_\nu$ (where $j_\nu = \rho u_\nu$ and $u_\nu = dx_\nu/d\tau$). In the traditional picture, this formulation applies the identical charge density for all components.

Provided that current-like charges are associated with q_T charges, and the real or Coulomb charges are q_V-denoted charges, we should apply $j_i = \rho_T u_i$ (i = 1, 2, 3), and $j_4 = ic\rho_V$ in the proposed new picture. This latter j_ν does not transform like a vector, and the electromagnetic force should be written as

$$F^{\mu} = F^{\mu\nu}\frac{1}{c}j_{\nu} = \begin{bmatrix} 0 & B_3 & -B_2 & -iE_1 \\ -B_3 & 0 & B_1 & -iE_2 \\ B_2 & -B_1 & 0 & -iE_3 \\ iE_1 & iE_2 & iE_3 & 0 \end{bmatrix} \begin{bmatrix} \rho_T\frac{u_1}{c} \\ \rho_T\frac{u_2}{c} \\ \rho_T\frac{u_3}{c} \\ i\rho_V \end{bmatrix} = \begin{bmatrix} B_3\rho_T\frac{u_2}{c} - B_2\rho_T\frac{u_3}{c} + E_1\rho_V \\ -B_3\rho_T\frac{u_1}{c} + B_1\rho_T\frac{u_3}{c} + E_2\rho_V \\ B_2\rho_T\frac{u_1}{c} - B_1\rho_T\frac{u_2}{c} + E_3\rho_V \\ iE_1\rho_T\frac{u_1}{c} + iE_2\rho_T\frac{u_2}{c} + iE_3\rho_T\frac{u_3}{c} \end{bmatrix} =$$

$$= \frac{1}{c}\begin{bmatrix} B_3u_2 - B_2u_3 & cE_1 \\ -B_3u_1 + B_1u_3 & cE_2 \\ B_2u_1 - B_1u_2 & cE_3 \\ iE_1u_1 + iE_2u_2 + iE_3u_3 & 0 \end{bmatrix} \begin{bmatrix} \rho_T \\ \rho_V \end{bmatrix} = \frac{1}{c}H^{\kappa l}\rho_l \tag{3.1}$$

where ($\kappa = 1, \ldots, 4$), ($l = 1, 2$), $E_i = -\partial_i\varphi - (1/c)(\partial A_i/\partial t)$, and $B_i = \mathrm{rot}_i\mathbf{A}$. It is easy to recognise that the first column of the matrix $H^{\kappa l}$ represents components of the Lorentz force, while the second column of the matrix represents components of the Coulomb force.

Here B_i are associated only with ρ_T, while E_i with both ρ_V and ρ_T, where $\varphi = \int(\rho_V/r)dV$ and $\mathbf{A} = \int[(j(\rho_T))/r]dV$ are the retarded scalar and vector potentials. It is obvious from the (3.1) matrix equation that this j_ν is *not* a four-vector, and ρ_V and ρ_T are mixed during the multiplication by $F^{\mu\nu}$; similarly, the components F^{μ} do not transform as vector components either.

Expression (3.1) is close to the approach applied by Dirac (1951a). The roots go back earlier. To see the origins, I must mention a few other historical steps.

I should mention that following the Symmetry Festival 2003 when the author first discussed the basic ideas – developed in detail in this book – with Yuval Ne'eman, there appeared a few similar approach publications. The roots are, however, much older. They are worth mentioning, because they lead to the definition (and justification of the *raison d'etre*) of a *velocity field* that we will intensively apply later in this book. Starting from the fundamental equation (Dirac, 1951a) $[\partial_\mu J^\mu A^\nu = 0]$, de Haas, E.P.J. (2004b), for example, derived similar (but not the same) conclusions like us, for QED and the SM, according to which physical real quantities can be derived by the distinction of the (spatially localised) electric potential and the Dirac velocity field. Although contrary to Dirac, our theory does not need to assume an ether; we can refer to his statement (Dirac, 1951b) where he defined the velocity field through the electromagnetic four-potential: "We have now the velocity at all points of space-time, playing a fundamental part in electrodynamics. It is natural to regard this as the velocity of some real physical thing." While Dirac identifies this "real physical thing" with an ether, our work is an attempt to identify these "things" with the quanta of a gauge field "localised" in that velocity field. For the author received objections since he first communicated the essence of the theory being presented in detail later in this book, what objections stated that the assumption of a velocity-dependent gauge contradicts localisation, I advise to keep in mind the cited words by Dirac (in addition to my main argument, the original formulation of Noether's second theorem, cf., Darvas, 2018c). De Haas assumes an analogy between Mie's (1912) non-gauge-invariant stress–energy tensor, and the stress–energy tensor in Dirac's 1951 theory in a four-velocity field. The analogy works only partially (in my opinion) – but the acknowledgement of the role of the velocity field in defining the stress–energy tensor is worth attention – since it partially confirms our approach, and leads to the same derivation of the Lorentz transformation of the electromagnetic field components, as we have interpreted it (Darvas, 2013b). As de Haas (2005) refers to it, the stress–energy tensor by M. von Laue (1911) can be written as $T_{\mu\nu}=J_{\mu}A_{\nu}$, where

$$A_{\nu}=\begin{bmatrix}\mathbf{A}\\ \frac{i}{c}\varphi\end{bmatrix}\text{ and } J_{\mu}=\begin{bmatrix}\mathbf{J}\\ ic\rho\end{bmatrix}$$

so

$$T_{\mu\nu}=\begin{bmatrix}\mathbf{J}\\ ic\rho\end{bmatrix}\begin{bmatrix}\mathbf{A}\\ \frac{i}{c}\varphi\end{bmatrix}=\begin{bmatrix}\mathbf{J}\otimes\mathbf{A} & \frac{i}{c}\varphi\mathbf{J}\\ ic\rho\mathbf{A} & -\rho\varphi\end{bmatrix}$$

which demonstrates the mentioned similitude to our approach. (Namely, the multiplication of two vectors – here one of them contains the charge-density current components, the other the field's vector and scalar potentials. One can check this by comparing them with the second column of $H^{\kappa l}$ in Eq. (3.1).) Nevertheless, our main goal here was not to find a full coincidence in the equations, rather the justification of the assumption, and preparing a discussion of a velocity field, which we will present in Section 6.

The result of the described (cf., eq. (3.1)) example is not in compliance with our experience! We lost some symmetry. As a consequence, in order to restore (Lorentz) invariance and compliance with experience, our programme must include requiring the existence of an additional transformation that should counteract the loss of symmetry caused by the introduction of two distinct states of the charges. This additional transformation will be specified in Section 5.

We will show later that the emerged problems demonstrated here in a classical electrodynamic example exist also in QED and appears also in field theoretical treatment of the distinction between field-charge pairs in general. Then, one of our main results in Section 5 of this book will be to derive the missing transformation that counteracts and replaces the demonstrated apparently lost symmetry in its role.

What is generally expected
is either a new symmetry principle
or possibly a new regime with
an altogether different set of
physical fields.
Gerard 't Hooft

Chapter II: **ISOTOPIC FIELD-CHARGES**

4 ISOTOPIC FIELD-CHARGES (IFC)

Some considerations similar to each other were formulated for the inertial and gravitational masses, and for the gravitational field on the one hand, as well as for the Coulomb and Lorentz charges, and the electromagnetic field on the other. Of course, the concrete forms are different for the gravitational field and the electromagnetic field, due to the essential difference between the energy–momentum tensor and the electromagnetic field tensor. Nevertheless, there are certain similarities concerning the application of the equivalence principle for the four-currents in both cases, and the split of the field-charges also in both cases. Identifying this invariance transformation (cf., Section 8), under the general covariance principle of nature, composes a part of the anticipated programme.

4.1 Field sources in the standard model (SM)

All fields are supposed to have sources. The sources of the individual fields are their charges. By consensus, the charges appearing in the scalar potential are considered the source of the given interaction field. We call the mass of gravity – that is, the source of the gravitational field – field-charge of the gravitational field. Similarly, we call the electric charge – that is, the source of the electromagnetic field – a field-charge of the electromagnetic field; flavour charges (of quarks and leptons) – that is, the sources of the weak field – field-charges of the weak field; the colour charges – that is, the sources of the strong field – field-charges of the strong field.

However, the mass of gravity is not the only field-charge of the gravitational field. Similar can be said about the charges of the other interaction fields.

These sources – field-charges – have been assumed to be realised in the matter field. They are, in the accustomed physical literature, considered also as sources for gauge fields.[4.1] Are these sources (of the two fields) really the same or should one distinguish two agents? We saw on the example of the mass that, in case of their transformation by velocity change, at least in the gravitational interaction, one should distinguish them. We saw on the example of the electric charge current that, in case of the electromagnetic interaction, one should distinguish them too. Later, we will investigate (cf., Section 9) the sources of the other interaction fields too. At his stage, let us restrict our scope of interest to the mass and the electric charge.

4.1 Partition of fields to matter and gauge fields is accepted in the physical literature. I disprefer this terminology. Sources of gauge fields, being either massive or massless quantities, are material substances as well. Therefore, I prefer to use other term pairs, where the respective name pairs are appropriate: scalar and vector fields, or potential and kinetic fields, and others.

https://doi.org/10.1515/9783110713183-004

We established similitude between the relation of gravitational mass and inertial mass (and their different physical roles), as well as of Coulomb charges and kinetic (Lorentz) charges. They are field-charges of different kind fields, however, both the gravitational mass and the electric Coulomb charge serve as sources of their respective fields, associated with a central, scalar potential, and both represent bound states of the respective field-charges. At the same time the inertial mass and the kinetic electric charge are both associated with the kinetic part of the Hamiltonian of their interaction field, both appear (as coefficients) in vector components of currents, as well as are the sources of momentum-like forces, namely the linear momentum and the magnetic momentum, respectively, and both represent free states of the respective field-charges. Provided that the assumption about the distinction between the members of the mass and electric charge twin siblings holds, they appear in the scalar and the vector parts of their Hamiltonians, respectively (and so do they in the Lagrangians of their interaction fields).

We considered the gravitational and inertial masses, as well as the Coulomb and Lorentz charges as twin pairs of each other, respectively. We saw that according to the covariance principle, the members of these pairs can be observed from various reference frames once as the potential and from another frame as the kinetic sibling of the pair. In certain physical situations, we cannot make a distinction between them. They appear like the same physical quantities in the standard physical equations. They behave in similar ways in almost all physical situations but at large velocities.

We indicated also that all *this distinction has a sense only if the two kinds of masses, as well as the two kinds of electric charges, can transform into each other.* Many signs showed that such a transformation should exist, otherwise they could not play once this and then that role in a physical system. We demonstrated the transformation of potential energy into kinetic energy in a classical mechanical example. We detected the phenomena; however, we have not seen any proof for the possibility of such a transformation between these siblings of the field sources. Therefore, we will look for the sought transformation in Sections 5 and 7.

4.2 Isotopic field-charges

4.2.1 Masses

Let us introduce the following definition: the mass of gravity and the mass of inertia will be considered as two equivalent quantity *isotopic states* of the field-charge of the gravitational field.

They represent two different qualities. The same massive object can behave once as a source of gravity, then as a measure of inertia. Their concepts express two properties of matter, whose existence was observed in different experiments. Physics established quantitative relations between them (i.e. equal values); however, this fact does not abolish their qualitative difference. We have all reason to make a distinction between them in our theories. Their different transformation rule justified that assumption.

For example, if you find in a grocery that buying a pound of apples and a pound of pears will cost you the same amount, you will neither mix them nor buy pears to flavour your birthday party's apple pie.

Mass of gravity is the coefficient of a scalar quantity, while the mass of inertia is the coefficient of a vector quantity with three (current) components as three projections in three independent spatial directions. The components of the four-momentum, composed with the help of the mass of inertia and the mass of gravity – compose a four-vector that are constituent parts of the equations of gravity in GTR.

In a frame of reference in rest, the quantity of the mass of gravity and the mass of inertia are measured equal. According to the equivalence principle their measures are proportional, and the factor of proportion was fixed as "1".

The mass of gravity is associated with the potential part V of the object's Hamiltonian. The mass of inertia is associated with the kinetic part T of the object's Hamiltonian. According to this observation, we named the mass of gravity as potential mass m_V, and the mass of inertia as kinetic mass m_T and call them together *isotopic field-charges (IFC) of the gravitational interaction.*

In GTR the inertial mass is attributed to the momentum densities, while the gravitational mass to the gravitational field energy. They are separated within the stress–energy tensor, but according to the general relativity principle they can be transformed into each other – we should add, at least in their quantitatively equivalent values. GTR does not make any statement about the qualitative transformation of the two kinds of masses into each other. This was a reason to identify them. The question simply has not emerged. Nevertheless, we will show that it cannot be avoided. More assumptions on the role of mass(es), especially in GTR, will be discussed partly in Section 10.2.1 and mainly in Section 11.2.

4.2.2 Electric charges

The Coulomb charge and the Lorentz charge are considered as two equivalent quantity *isotopic states* of the field-charge of the electromagnetic field. The asymmetric role played by two interacting electric charges was revealed in the Møller scattering. Its consequences cannot be avoided in many physical phenomena. The Coulomb charge is the source of a static electric force, while the Lorentz charge is the source of a kinetic, magnetic force. They represent two different qualities. The same electrically charged object can behave once as a source of a static electric field, then as a source of a magnetic field. Their concepts express two properties of electromagnetic charges, whose existence was observed in different experiments. Electrodynamics established quantitative relations between them, but this revealed equal value does not abolish their qualitative difference. We have all reason to make a similar distinction between them in our theories like we did in the case of the masses. Their different effects (electrostatic and magnetic) justified that assumption.

Electric (Coulomb) charge is the coefficient of a scalar quantity, while the current-like Lorentz charge is the coefficient of a vector quantity with three components as projections in three independent spatial directions. The components of the four-momentum – composed with the help of the Coulomb and Lorentz charges – compose a vierbein and they are constituent parts of the electromagnetic equations in quantum electrodynamics (QED).

In a given frame of reference, the quantity of the electric Coulomb charge and the Lorentz charge as the source of magnetism are measured equal. According to the definition of equivalence principle for electric charges (Section 3.3), their measures are proportional, and the factor of proportion is fixed as “1”.

The Coulomb charge is associated with the potential (scalar) part of the object's Hamiltonian. The Lorentz charge is associated with the kinetic (vector) part of the object's Hamiltonian. Thus, we named the Coulomb charge as scalar charge q_V, and the Lorentz charge as kinetic charge q_T and call them together as *IFC of the electromagnetic interaction.* In QED the kinetic charge q_T is attributed to the magnetic field components, while the scalar charge q_V to the Coulomb field's source. They are separated within a system's field components, but according to the general covariance principle they can be transformed into each other – we should add, at least in their quantitatively equivalent values.

Consequently, there are two qualitatively different proportional factors belonging to the potential and kinetic energies of an individual object both in gravitational and in electromagnetic interaction fields. Maintaining that they quantitatively coincide, respectively, we should reflect their qualitative difference in our equations of motion. (Similar to the greengrocer shopkeeper, who books the sold equal-priced pears and apples and their stocks separated.) At least, until there appears a velocity boost (or a shock in the pear market).

For the number of letters in the Latin and Greek alphabets is limited, we denote the *IFC* of any field by a common symbol, the fourth letter of the Hebrew alphabet, dalet (ד) (Darvas, 2011).

4.3 The identity-equivalence diversity on the example of the isotopic spin

To generalise the identity-equivalence dispute, we extend its discussion beyond gravity and electromagnetism. The distinction between the proton and the neutron mentioned in Section 2.2 exposes an already classical interpretation of the identity-equivalence problem. We quote the example of the isotopic spin because it will play an analogical role in our further discussion in this book.

Classically the proton and the neutron were considered as two different particles, which made up the nucleus of the atoms. They were (approximately) equivalent in all of their properties but differed in their electric charge. Later, there was a period, when their role in weak interaction was emphasized, in which, in course of negative beta decay, a neutron spontaneously emits an electron (and an antineutrino) and transforms into a proton. According to the SM (standard model) of physics, they are considered as two particles with internal structure, which are made up of *u* and *d* quarks. The difference is that the proton is made up of two *u* and one *d* quark (*udu*), while the neutron of two *d* and one *u* quark (*dud*). All these interpretations considered the proton and the neutron as two different particles.

There was also a fourth approach to them since almost beginning in the 1930s. The background of this picture was that the only property that distinguishes a proton from a neutron (within a nucleus) is their different electric charge. Charge is the source of the electromagnetic field, and thus it is subject of electromagnetic interaction. However, if electromagnetic interaction would prevail within the nucleus, the nucleus would have to explode, due to the repulsive electrostatic forces between the identically positive charged *u* quarks. This does not happen. Consequently, it was assumed that there must be a strong force, which effects in the range of the diameter of the nucleus, and not only counterbalances the repulsive electromagnetic force but also even keeps the nucleons stable together. In the range where strong force suppresses the effect of electromagnetic forces, one cannot observe any difference between electrically charged constituents that is caused only by electromagnetic properties. We saw, proton and neutron differ in this property only. Therefore they are indistinguishable within the nucleus. Consequently, this approach states, that the proton and the neutron are "identical" (?) particles: nucleons, only they represent two "states of the nucleon". Is this a play with the words or something more?

So, are they identical or different? Like a black and a white cat in a dark room, one cannot make a distinction between them. At least in those conditions, they are observed as identical. Their symmetry manifested itself in the fact that one does not experience any difference if the two cats are swapped with each other. As soon as the conditions change, (one turns the light in the cats' room on, or detects nucleons outside an atom) the previously suppressed properties happen to play an active role. We say that the "identity", that means, interchangeability of the proton and the neutron under the conditions of strong interaction inside a nucleus, is an improper symmetry. It is improper because it is valid only among those conditions. We say also that the symmetry is broken under other conditions. In fact, they can be considered identical under conditions while they play equivalent roles in their system and until they are indistinguishable for the observer. As soon as those circumstances are changed, they lose that conditional identity.

To describe this symmetry, it was proposed that the proton and the neutron be considered "identical" particles that can take two states of a property called *isotopic spin*, or shortly isospin. This name expresses that they represent two observed *isotopic states* of an identical particle (like bullets shot by two guns and flying once with a right-handed, then with a left-handed rotation). The distinction is similar to the spin states of any other particle, which are also considered identical. For example, two electrons in "up" and "down" spin states are both considered electrons, and not particles denoted by different names. Unlike the spin, which is a property (own angular momentum) interpreted in the space, the analogously introduced isotopic spin property presumes the existence of an abstract (gauge) field, where it can be interpreted.

It was proven (Yang and Mills, 1954) that the total isotopic spin is a conserved property (at that time at least) in nucleon–nucleon interactions, and later it was shown that all strong interactions must satisfy the same conservation law. The conservation of isotopic spin is identical with the requirement of invariance of all interactions under isotopic spin rotation (a symbolic "rotation" in the gauge field where it is interpreted). This means that when electromagnetic interactions can be disregarded, the orientation of the isotopic spin is of no physical significance. For this reason, an isotopic gauge (later called Yang–Mills gauge after the discoverers) was introduced as an arbitrary way of choosing the orientation of the isotopic spin axes at all space–time points, in analogy with the electromagnetic gauge which represents an arbitrary way of choosing the complex phase factor of a charged field at all space–time points, and which latter had led to the proof of the charge conservation. (Arbitrariness is meant here with the limitation, that once one chooses what isotopic state to call a proton and what a neutron at one space–time point, one is then no more free to make any choice at other space–time points.) In short, all physical interactions (that do not involve the electromagnetic field) are invariant under a transformation in an isotopic gauge field, that is, under a space–time dependent rotation of the property, called isotopic spin, which can take two stable eigenvalues.

Considering the example of a Deuterium nucleus that consists of a proton and a neutron, we can rather say that it consists of two identical particles which can be found in one of two isotopic spin states both. One of them can be found at p_1 probability in isospin state up, and p_2 probability in isospin state down. At the same moment, the other can be found at p_2 probability in isospin state up, and p_1 probability in isospin state down. The sum of the two probabilities should amount to 1 because certainly one of them takes the up and the other the down isospin state in the isotopic spin field.

Similar to the spin, the rotation of the isotopic spin, in its own field, is subject to an SU(2) symmetry. Later a similar property of the weakly interacting particles, subject also to SU(2) symmetry, was found. Due to the analogy of their transformation rule, this property got the name weak isospin. In the SM, we had altogether three properties subject to SU(2) symmetry: the spin, the isotopic spin, and the weak isospin. All the three can rotate in their own field.

The field-charges are also indistinguishable among the (energetic) conditions prevailing in the SM. At ultrahigh kinetic energies, the difference between their two states becomes apparent. Their state function splits in two. The example of broken symmetries of the isotopic spin and the weak isospin justifies to call them also isotopic states of a quantity (namely of the field-charges). However, their transformation rules should be still examined. This will be discussed in Section 8.

4.4 3 + 1 quantities in physics

We showed that IFC ($\daleth_{\mathrm{V}}$ and $\daleth_{\mathrm{T}}$) of the gravitational and the electromagnetic interaction fields appear in the physical formulas in split forms: one of them associated with a three-vector and the other with a scalar. Therefore, they appear in 3 + 1 times in each concrete instance.

There are several 3 + 1 parameter quantities in physics (like space–time, vector + scalar potentials, four-currents, and four-momentum). In most cases (but space–time), the three- and the one-parameter characterised elements of these quantities differ in the field sources (e.g. inertial and gravitational masses, and Lorentz- and Coulomb-type electric charges) associated with them. The members of the field-source pairs appear in the vector and the scalar potentials, respectively.

The field sources assigned to the vector quantities coincide in all the three vector components. (A simple example of the mechanical four-momentum: its components look like $m_T v_x$, $m_T v_y$, and $m_T v_z$; while the fourth component is defined as $m_V c$, where v_i are velocity components, c the speed of light, and m_T and m_V are the inertial and gravitational masses, respectively.) We describe an algebra that can handle such (3 + 1) type of quantities.

We will describe a (3 + 1) parameter algebra in Section 5 of this book. It will be called τ algebra. The name is justified by the Dirac algebra. Dirac defined his γ matrices by the help of his ρ and Pauli's σ matrices. We will apply certain analogies to the Dirac theory, so we use the next letter, τ of the Greek alphabet. Then, using the analogy of the Dirac algebra, we will define matrices called δ, similar to the γ matrices.

5 HYPERSYMMETRY[5.1] (HySy)

5.1 Matrix algebra for 3 + 1 parametric transformations

Isotopic field-charge (IFC) theory takes into account sources of respective scalar and vector fields as different physical properties. They are considered as isotopic states of each other that are, at least at rest, quantitatively equivalent, but qualitatively different. Putting them in the formula of field-charge currents, they would destroy the invariance of those under usual transformations, unless an additional invariance that can transform the two isotopic states into each other restores the "lost" order. The present section explores the algebra of this invariance and the respective group that transforms the isotopic states of field-charges of any fundamental physical interaction field into each other.

The IFC as sources of vector fields appear in the field-charge currents as multipliers of three-velocities, therefore the algebra of their transformation into IFC of the respective scalar field, and *vice versa*, should take the form of 4×4 matrices, in which there must appear three-row minors of coinciding elements that act on the IFC multiplying the three velocity vector components.

Since we will apply the algebra to be described in this section – as an example – for an extended form of "the" original Dirac equation, we need to introduce partly the basics and partly a minor extension to the Dirac algebra.

5.1.1 Vector algebra and quaternion algebra

Vectors can be transformed into quaternions: multiplying them by the imaginary unit, and *vice versa*. When one substitutes vectors with quaternions or quaternions with vectors obtained in this way in physical equations, they will not change the physical meaning of the equations. In other words, they do not bring in extra physical information. (This statement does not concern certainly those quaternions and vectors that have not been obtained in this way.)

We use the following notations:

$$\mathbf{i} = \mathbf{j} = \mathbf{k} \text{ vectors;} \quad \boldsymbol{i} = \boldsymbol{j} = \boldsymbol{k} \text{ quaternions,} \quad \mathbf{i} = i\boldsymbol{i}.$$

Vectors are subjects of the following algebra:

$$\mathbf{i}^2 = \mathbf{j}^2 = \mathbf{k}^2 = 1; \ \mathbf{ijk} = i$$

5.1 The term *hypersymmetry* was used in physics last in the mid-1990s, although in quite another meaning. Its appearance was related to the childhood of supersymmetry in the meaning of an extension, in other publications as generalisation of that (first of all by R. Kerner et al.). That meaning was abandoned in short and the proposed algebra was not used for long. We are applying this term in a new meaning for an alternative, competing model of the supersymmetry.

https://doi.org/10.1515/9783110713183-005

anticommutation rules: $\{\mathbf{i}, \mathbf{j}\} = \{\mathbf{j}, \mathbf{k}\} = \{\mathbf{k}, \mathbf{i}\} = 0$
commutation rules: $[\mathbf{i}, \mathbf{j}] = 2i\mathbf{k}, \quad [\mathbf{j}, \mathbf{k}] = 2i\mathbf{i}, \quad [\mathbf{k}, \mathbf{i}] = 2i\mathbf{j}$

Quaternions are subjects of the following algebra:

$$\boldsymbol{i}^2 = (-i)^2\mathbf{i}^2 = -1 \quad \boldsymbol{j}^2 = (-i)^2\mathbf{j}^2 = -1 \quad \boldsymbol{k}^2 = (-i)^2\mathbf{k}^2 = -1 \quad \boldsymbol{ijk} = -1$$

anticommutation rules: $\{\boldsymbol{i}, \boldsymbol{j}\} = \{\boldsymbol{j}, \boldsymbol{k}\} = \{\boldsymbol{k}, \boldsymbol{i}\} = 0$
commutation rules: $[\boldsymbol{i}, \boldsymbol{j}] = 2\boldsymbol{k}, \ [\boldsymbol{j}, \boldsymbol{k}] = 2\boldsymbol{i}, \ [\boldsymbol{k}, \boldsymbol{i}] = 2\boldsymbol{j}$.

Several statements mentioned earlier continue certain ideas outlined in recent years' publications by P. Rowlands (2012a, 2013a, 2015a, b).

Let us apply this algebra to Pauli's and Dirac's matrices with the following notations:

$$\begin{array}{lll} \mathbf{i} = \boldsymbol{\sigma}_1 & \mathbf{j} = \boldsymbol{\sigma}_2 & \mathbf{k} = \boldsymbol{\sigma}_3 \\ \boldsymbol{i} = -i\mathbf{i} = -i\boldsymbol{\sigma}_1 & \boldsymbol{j} = -i\mathbf{j} = -i\boldsymbol{\sigma}_2 & \boldsymbol{k} = -i\mathbf{k} = -i\boldsymbol{\sigma}_3 \\ \boldsymbol{\sigma}_i = -i\boldsymbol{\sigma}_i & \boldsymbol{\sigma}_i = i\boldsymbol{\sigma}_i & (\mathbf{i}, \mathbf{j}, \mathbf{k}) \\ \boldsymbol{\rho}_i = -i\boldsymbol{\rho}_i & \boldsymbol{\rho}_i = i\boldsymbol{\rho}_\mathbf{i} & \end{array}$$

Using these definitions, we can interpret Dirac's matrices as double vector algebra (presented in vector and quaternion combinations). They can be expressed as products of ρ and σ matrices. Dirac (1928) first introduced matrices marked α, then the matrices γ, named after him:

$$\boldsymbol{\alpha}_i = \boldsymbol{\rho}_1\boldsymbol{\sigma}_i = i\boldsymbol{\rho}_1\boldsymbol{\sigma}_i = i\boldsymbol{\rho}_1\boldsymbol{\sigma}_i = -\boldsymbol{\rho}_1\boldsymbol{\sigma}_i$$

$$\boldsymbol{\gamma}_i = \boldsymbol{\rho}_2\boldsymbol{\sigma}_i = i\boldsymbol{\rho}_2\boldsymbol{\sigma}_i = i\boldsymbol{\rho}_2\boldsymbol{\sigma}_i = -\boldsymbol{\rho}_2\boldsymbol{\sigma}_i$$

Observe, that the $\boldsymbol{\alpha}$ and γ matrices are generated by the help of the ρ_1 and ρ_2 matrices as coefficients of the σ_i matrices. We extend this set of matrices with matrices to be called β, generated by the help of ρ_3 (Darvas, 2015b):

$$\boldsymbol{\beta}_i = \boldsymbol{\rho}_3\boldsymbol{\sigma}_i = i\boldsymbol{\rho}_3\boldsymbol{\sigma}_i = i\boldsymbol{\rho}_3\boldsymbol{\sigma}_i = -\boldsymbol{\rho}_3\boldsymbol{\sigma}_i$$

The latter did not appear in the system of matrices introduced by Dirac. Notice that all the mentioned product combinations of either vectors, quaternions, or both provide vectors.

The vector and quaternion algebra of all the earlier listed five types of matrices are summarised in the following table:

$\boldsymbol{\rho}_i$:	$\boldsymbol{\rho}^2{}_i = 1$	$\boldsymbol{\rho}_i\boldsymbol{\rho}_j = -\boldsymbol{\rho}_j\boldsymbol{\rho}_i$	$\boldsymbol{\rho}_i\boldsymbol{\rho}_j = i\boldsymbol{\rho}_k$	$\boldsymbol{\rho}_i = -i\boldsymbol{\rho}_i$	$\boldsymbol{\rho}_i^{\ 2} = -1$	$\boldsymbol{\rho}_i\boldsymbol{\rho}_j = -\boldsymbol{\rho}_j\boldsymbol{\rho}_i$	$\boldsymbol{\rho}_i\boldsymbol{\rho}_j = \boldsymbol{\rho}_k$
$\boldsymbol{\sigma}_i$:	$\boldsymbol{\sigma}_i^{\ 2} = 1$	$\boldsymbol{\sigma}_i\boldsymbol{\sigma}_j = -\boldsymbol{\sigma}_j\boldsymbol{\sigma}_i$	$\boldsymbol{\sigma}_i\boldsymbol{\sigma}_j = i\boldsymbol{\sigma}_k$	$\boldsymbol{\sigma}_i = -i\boldsymbol{\sigma}_i$	$\boldsymbol{\sigma}_i^{\ 2} = -1$	$\boldsymbol{\sigma}_i\boldsymbol{\sigma}_j = -\boldsymbol{\sigma}_j\boldsymbol{\sigma}_i$	$\boldsymbol{\sigma}_i\boldsymbol{\sigma}_j = \boldsymbol{\sigma}_k$
$\boldsymbol{\alpha}_i = \boldsymbol{\rho}_1\boldsymbol{\sigma}_i$	$\boldsymbol{\alpha}_i^{\ 2} = 1$	$\boldsymbol{\alpha}_i\boldsymbol{\alpha}_j = -\boldsymbol{\alpha}_j\boldsymbol{\alpha}_i$	$\boldsymbol{\alpha}_i\boldsymbol{\alpha}_j = i\boldsymbol{\sigma}_k$	$\boldsymbol{\alpha}_i = -i\boldsymbol{\alpha}_i$	$\boldsymbol{\alpha}_i^{\ 2} = -1$	$\boldsymbol{\alpha}_i\boldsymbol{\alpha}_j = -\boldsymbol{\alpha}_j\boldsymbol{\alpha}_i$	$\boldsymbol{\alpha}_i\boldsymbol{\alpha}_j = \boldsymbol{\sigma}_k$
$\boldsymbol{\gamma}_i = \boldsymbol{\rho}_2\boldsymbol{\sigma}_i$	$\boldsymbol{\gamma}_i^{\ 2} = 1$	$\boldsymbol{\gamma}_i\boldsymbol{\gamma}_j = -\boldsymbol{\gamma}_j\boldsymbol{\gamma}_i$	$\boldsymbol{\gamma}_i\boldsymbol{\gamma}_j = i\boldsymbol{\sigma}_k$	$\boldsymbol{\gamma}_i = -i\boldsymbol{\gamma}_i$	$\boldsymbol{\gamma}_i^{\ 2} = -1$	$\boldsymbol{\gamma}_i\boldsymbol{\gamma}_j = -\boldsymbol{\gamma}_j\boldsymbol{\gamma}_i$	$\boldsymbol{\gamma}_i\boldsymbol{\gamma}_j = \boldsymbol{\sigma}_k$
$\boldsymbol{\beta}_i = \boldsymbol{\rho}_3\boldsymbol{\sigma}_i$	$\boldsymbol{\beta}_i^{\ 2} = 1$	$\boldsymbol{\beta}_i\boldsymbol{\beta}_j = -\boldsymbol{\beta}_j\boldsymbol{\beta}_i$	$\boldsymbol{\beta}_i\boldsymbol{\beta}_j = i\boldsymbol{\sigma}_k$	$\boldsymbol{\beta}_i = -i\boldsymbol{\beta}_i$	$\boldsymbol{\beta}_i^{\ 2} = -1$	$\boldsymbol{\beta}_i\boldsymbol{\beta}_j = -\boldsymbol{\beta}_j\boldsymbol{\beta}_i$	$\boldsymbol{\beta}_i\boldsymbol{\beta}_j = \boldsymbol{\sigma}_k$

The vectors and quaternions β were introduced by the author to make the table complete. We must note that the vectors and quaternions appearing in the earlier table are not all independent. These five matrices can be expressed by two of the others (but not in all arbitrary combinations). $\boldsymbol{\sigma}$ plays a distinguished role, which restricts the number of allowed representations. The different pairings provide different dual representations of the Dirac algebra. We see that they can be represented both in vector and quaternion (pseudovector) forms.

Note, that $\boldsymbol{\alpha}$, $\boldsymbol{\gamma}$, and $\boldsymbol{\beta}$ do not transform according to the **i**, **j**, **k** vector algebra, like $\boldsymbol{\rho}$ and $\boldsymbol{\sigma}$ do. Their commutators are not produced from their own set, but they are a component of $\boldsymbol{\sigma}$.

The roles of $\boldsymbol{\rho}$ and $\boldsymbol{\sigma}$ can be changed. This allows an interesting observation and leads to a third (mixed) algebra. Nevertheless, this allows another set of representations.

For $\boldsymbol{\rho}$ and $\boldsymbol{\sigma}$ commute, $\boldsymbol{\rho}_i\boldsymbol{\sigma}_j = \boldsymbol{\rho}_j\boldsymbol{\sigma}_i$, let us compose the following combinations:

$$\begin{array}{lll}
\boldsymbol{\rho}_1\boldsymbol{\sigma}_1 = \boldsymbol{\sigma}_1\boldsymbol{\rho}_1 = \boldsymbol{\alpha}_1 = \boldsymbol{\kappa}_1 & \boldsymbol{\rho}_2\boldsymbol{\sigma}_1 = \boldsymbol{\sigma}_1\boldsymbol{\rho}_2 = \boldsymbol{\gamma}_1 = \boldsymbol{\kappa}_2 & \boldsymbol{\rho}_3\boldsymbol{\sigma}_1 = \boldsymbol{\sigma}_1\boldsymbol{\rho}_3 = \boldsymbol{\beta}_1 = \boldsymbol{\kappa}_3 \\
\boldsymbol{\rho}_1\boldsymbol{\sigma}_2 = \boldsymbol{\sigma}_2\boldsymbol{\rho}_1 = \boldsymbol{\alpha}_2 = \boldsymbol{\lambda}_1 & \boldsymbol{\rho}_2\boldsymbol{\sigma}_2 = \boldsymbol{\sigma}_2\boldsymbol{\rho}_1 = \boldsymbol{\gamma}_2 = \boldsymbol{\lambda}_2 & \boldsymbol{\rho}_3\boldsymbol{\sigma}_2 = \boldsymbol{\sigma}_2\boldsymbol{\rho}_3 = \boldsymbol{\beta}_2 = \boldsymbol{\lambda}_3 \\
\boldsymbol{\rho}_1\boldsymbol{\sigma}_3 = \boldsymbol{\sigma}_3\boldsymbol{\rho}_1 = \boldsymbol{\alpha}_3 = \boldsymbol{\mu}_1 & \boldsymbol{\rho}_2\boldsymbol{\sigma}_3 = \boldsymbol{\sigma}_3\boldsymbol{\rho}_1 = \boldsymbol{\gamma}_3 = \boldsymbol{\mu}_2 & \boldsymbol{\rho}_3\boldsymbol{\sigma}_3 = \boldsymbol{\sigma}_3\boldsymbol{\rho}_3 = \boldsymbol{\beta}_3 = \boldsymbol{\mu}_3
\end{array}$$

These equalities serve as definitions of the κ, λ, and μ matrices. κ, λ, μ are defined by a new indexation (row–column change); they provide no new matrices, however, allow to compose new vectors, in the following way:

$$\boldsymbol{\sigma}_1\boldsymbol{\rho}_i = \boldsymbol{\kappa}_i \quad \boldsymbol{\sigma}_2\boldsymbol{\rho}_i = \boldsymbol{\lambda}_i \quad \boldsymbol{\sigma}_3\boldsymbol{\rho}_i = \boldsymbol{\mu}_i$$

$$\begin{pmatrix} \alpha_1 \\ \gamma_1 \\ \beta_1 \end{pmatrix} = \boldsymbol{\kappa}_i; \qquad \begin{pmatrix} \alpha_2 \\ \gamma_2 \\ \beta_2 \end{pmatrix} = \boldsymbol{\lambda}_i; \qquad \begin{pmatrix} \alpha_3 \\ \gamma_3 \\ \beta_3 \end{pmatrix} = \boldsymbol{\mu}_i$$

They follow the algebra:

$$\boldsymbol{\kappa}_i^2 = \boldsymbol{\lambda}_i^2 = \boldsymbol{\mu}_i^2 = 1$$

$\boldsymbol{\kappa}_i\boldsymbol{\kappa}_j = -\boldsymbol{\kappa}_j\boldsymbol{\kappa}_i \quad \boldsymbol{\kappa}_i\boldsymbol{\kappa}_j = i\boldsymbol{\rho}_k$ (and the same rules apply for $\boldsymbol{\lambda}$ and $\boldsymbol{\mu}$ too).

Expressing the similar equalities with quaternions:

$$\boldsymbol{\kappa}_i = -i\boldsymbol{\kappa}_i; \qquad \boldsymbol{\lambda}_i = -i\boldsymbol{\lambda}_i; \qquad \boldsymbol{\mu}_i = -i\boldsymbol{\mu}_i$$

$$\boldsymbol{\kappa}_i^2 = \boldsymbol{\lambda}_i^2 = \boldsymbol{\mu}_i^2 = -1$$

$\boldsymbol{\kappa}_i\boldsymbol{\kappa}_j = -\boldsymbol{\kappa}_j\boldsymbol{\kappa}_i \quad \boldsymbol{\kappa}_i\boldsymbol{\kappa}_j = \boldsymbol{\rho}_k$ (and the same rules apply for $\boldsymbol{\lambda}$ and $\boldsymbol{\mu}$ too)

The $\boldsymbol{\kappa}, \boldsymbol{\lambda}, \boldsymbol{\mu}$ representation ensures a $\boldsymbol{\sigma}$-$\boldsymbol{\rho}$ mirror-symmetric algebra of the $\boldsymbol{\alpha}, \gamma, \boldsymbol{\beta}$ representation. While the generators of the $\boldsymbol{\alpha}, \gamma, \boldsymbol{\beta}$ double vector representation were $\boldsymbol{\rho}$, the generators of the $\boldsymbol{\kappa}, \boldsymbol{\lambda}, \boldsymbol{\mu}$ double vector representation are $\boldsymbol{\sigma}$.

Let us re-cluster the $\boldsymbol{\kappa}, \boldsymbol{\lambda}, \boldsymbol{\mu}$ matrices. They can define another (mixed, vector–quaternion-like) Clifford algebra. They define a semi-vector – semi-quaternion algebra. The κ, λ, μ matrices

- commute
- square, like quaternions
- multiply, like vectors

$$\boldsymbol{\kappa}_i^2 = \boldsymbol{\lambda}_i^2 = \boldsymbol{\mu}_i^2 = 1 \quad \boldsymbol{\kappa}_i\boldsymbol{\lambda}_j = \boldsymbol{\lambda}_j\boldsymbol{\kappa}_i \quad \boldsymbol{\kappa}_i\boldsymbol{\lambda}_j = -\boldsymbol{\mu}_k$$

$$\boldsymbol{\kappa}_\mathrm{i}^2 = \boldsymbol{\lambda}_\mathrm{i}^2 = \boldsymbol{\mu}_\mathrm{i}^2 = -1 \quad \boldsymbol{\kappa}_\mathrm{i}\boldsymbol{\lambda}_\mathrm{j} = \boldsymbol{\lambda}_\mathrm{j}\boldsymbol{\kappa}_\mathrm{i} \quad \boldsymbol{\kappa}_\mathrm{i}\boldsymbol{\lambda}_\mathrm{j} = \mathrm{i}\boldsymbol{\mu}_\mathrm{k}$$

5.1.2 Vector and quaternion algebras applied to physics

All the representations in Section 5.1.1 indicate Clifford algebras generated either by vectors or by quaternions over a spinor field. Their specifics are that the roles of the applied vectors or quaternions or their mixtures (in the doublets) can be changed and may serve once as generators of a *field*, once as bases of a (vector, quaternion) *space*. Any of them may serve as a generator – at least in formal mathematical terms – however, it is doubtful, at least with our present-day knowledge of physics, whether a definite physical meaning can be assigned to all (formally) so generated fields. (As Wigner noticed in his famous essay on the unreasonable effectiveness of mathematics in the sciences, mathematics allows more than nature can realise. This can be interpreted as underdetermination seen from one side and overdetermination seen from the other.)

5.2 The algebra of hypersymmetry (HySy)

5.2.1 Introduction to the τ-algebra

We investigate matrices (that we denote by τ) that can function as operators. We need this latter condition in order to apply the findings to a quantum mechanical way of description of physical phenomena. Operators affect state functions. One can define eigenvalue equations in which the effect of the τ operators is multiplication by numbers; such numbers are called eigenvalues of the τ operator. The state functions in these eigenvalue equations take the form of eigenfunctions of the τ

operators. Since we want to represent the linear operator τ in the form of square matrices, we are looking for the eigenfunctions in the form of eigenvectors.

In usual quantum mechanical procedures, one knows the operators and looks for state functions (eigenfunctions) which the operators act to provide eigenvalues. We proceed oppositely. We are looking for operators (in matrix form) whose eigenfunctions are known for us. Namely, we would like to find such τ operators that are able to transform vierbeins characterised by 3+1 parameters into each other (Vaccaro, 2016).[5.2] The operators τ should fulfil eigenvalue equations $\tau\varphi=k\varphi$, where φ are eigenfunctions of the operator τ, and k are numbers, eigenvalues of this equation. As we will show in *Theorem 4*, the tau algebra will be unitary, and we will find (Section 5.2.4) the eigenvalues of τ to be ±½. We denote the τ operators' eigenfunctions belonging to the eigenvalues of τ by $\varphi_+^{(\tau)}$ and $\varphi_-^{(\tau)}$, respectively. For simplicity, we denote the vierbein forms of the latter two eigenvectors by χ and ϑ, respectively (cf., e.g. Eq. (5.13)). Having known the properties of the expected eigenvalues and the eigenfunctions, we are seeking operators τ that satisfy the eigenequation $\tau\varphi=k\varphi$.

5.2.2 The τ-algebra

Let us choose two eigenvectors χ and ϑ [with (+ + + −) signature] in the following representation:

$$\chi=\begin{bmatrix}1\\1\\1\\0\end{bmatrix},\qquad \vartheta=\begin{bmatrix}0\\0\\0\\i\end{bmatrix} \tag{5.1}$$

Note that we had certain limited freedom to choose the values in the eigenvectors. This choice makes handling the τ transformation-matrices convenient. Recall, that in the following case, like in many other cases, the representation of the transformation group coincides with a representation of the respective Lie algebra.

We aim at finding matrices, which transform linearly these two eigenvectors – that satisfy the eigenvalue equations for the operator τ – as described later. Similar

5.2 In contrast to the Dirac bispinor consisting of 2+2 components (two-spinors), our spinors have a peculiarity that must be reflected in their construction. Since they are expected to make (in classical terms unusual) correspondence between *vector components* and *scalars* (c.f., also Sections 5.2.3 and 5.2.5), they must be constructed in a 3+1 form. Four-columns of these bispinors can be subjects of linear transformation by [4 × 4] matrices whose minors should reflect that 3+1 structure. This predicts the structure and explains the character of the sought τ-matrices.

to the σ algebra in Dirac's (1928) quantum electrodynamics (QED), we are looking for matrices τ with the following properties:

$$\tau_3\chi = \chi \qquad \tau_3\vartheta = -\vartheta$$

$$\tau_2\chi = \vartheta \qquad \tau_2\vartheta = \chi$$

$$\text{requiring } \tau_i\tau_j = i\tau_k \quad (i,\ j,\ k \text{ cyclic indices}):$$

$$\tau_1\chi = -i\vartheta \qquad \tau_1\vartheta = i\chi$$

A set of the following matrices meets the required properties:

$$\tau_1 = \begin{bmatrix} 0 & 0 & 0 & 1 \\ 0 & 0 & 0 & 1 \\ 0 & 0 & 0 & 1 \\ 1 & 0 & 0 & 0 \end{bmatrix}, \quad \tau_2 = \begin{bmatrix} 0 & 0 & 0 & -i \\ 0 & 0 & 0 & -i \\ 0 & 0 & 0 & -i \\ i & 0 & 0 & 0 \end{bmatrix}, \quad \tau_3 = \begin{bmatrix} 1 & 0 & 0 & 0 \\ 1 & 0 & 0 & 0 \\ 1 & 0 & 0 & 0 \\ 0 & 0 & 0 & -1 \end{bmatrix} \tag{5.2}$$

These τ-matrices satisfy the following algebra:

$$\tau_1^2 = \tau_2^2 = \tau_3^2 = \begin{bmatrix} 1 & 0 & 0 & 0 \\ 1 & 0 & 0 & 0 \\ 1 & 0 & 0 & 0 \\ 0 & 0 & 0 & 1 \end{bmatrix} = \mathbf{E},\ \{\tau_i, \tau_j\} = 0,\ [\tau_i, \tau_j] = 2i\tau_k \tag{5.3}$$

Introducing the notation: $\begin{bmatrix} 1 & 0 & 0 \\ 1 & 0 & 0 \\ 1 & 0 & 0 \end{bmatrix} = \mathbf{I}_L,\ \begin{bmatrix} 0 & 0 & 1 \\ 0 & 0 & 1 \\ 0 & 0 & 1 \end{bmatrix} = \mathbf{I}_R$ the (5.2) matrices can be written in the form: $\tau_1 = \begin{bmatrix} 0 & \mathbf{I}_R \\ 1 & 0 \end{bmatrix}, \tau_2 = \begin{bmatrix} 0 & -i\mathbf{I}_R \\ i & 0 \end{bmatrix}, \tau_3 = \begin{bmatrix} \mathbf{I}_L & 0 \\ 0 & -1 \end{bmatrix}$

where

$$\tau_1^2 = \tau_2^2 = \tau_3^2 = \mathbf{E} = \begin{bmatrix} \mathbf{I}_L & 0 \\ 0 & 1 \end{bmatrix}. \tag{5.4}$$

We denote the (diagonal) identity matrix of $[4 \times 4]$ matrices by [**1**]. The *unit matrix* of the τ matrix-algebra is denoted by **E** [cf., eq. (5.4)].

5.2.3 Group properties of the τ-matrices

Theorem 1: The τ-matrices generate a group.

Proof: The τ-matrices satisfy the four group axioms. The products of the τ_i, plus the unit matrix **E**, together with their conjugate and negated matrices form a closed group; they are associative; there exists an identity element of the *group*, which is the unit matrix **E** defined by (5.3); and since the squares of all τ_i produce the unit matrix, all τ_i are the inverse group elements of themselves, that means $\tau_i^{-1} = \tau_i$. Note that the identity element of the group in a group-theoretical sense is not the identity matrix [**1**], instead, it is the unit matrix **E** defined in (5.4).

□

This is the group of HySy. Representation of the transformation group is determined by the algebra of the τ-matrices.

(As an example, they transform the two states of the IFC among each other by a rotation in an abstract field described by Darvas [2011]. This is just what we wanted to show and indicated in the introductory sections.)

The expressions (5.4) of the τ-matrices show a very similar form to the [2 × 2] Pauli matrices:

$$\sigma_1 = \begin{bmatrix} 0 & 1 \\ 1 & 0 \end{bmatrix}, \quad \sigma_2 = \begin{bmatrix} 0 & -i \\ i & 0 \end{bmatrix}, \quad \sigma_3 = \begin{bmatrix} 1 & 0 \\ 0 & -1 \end{bmatrix} \tag{5.5}$$

Later, we will use the partial formal analogy in Section 5.3. However, there are differences as well: in contrast to the diagonal σ matrices, our τ-matrices manifest an apparent chiral character.

Observe that all the earlier matrices of the τ-algebra are singular. In matrix algebraic terms, in general, singular matrices have neither inverse nor adjoint matrices! This limits their commutation in scalar multiplication with other matrices. Nevertheless, in a group-theoretical sense, they have inverse and unit elements, considering the expressions in (5.4). Inverse elements of the singular matrices constituting the group are interpreted in Penrose–Moore sense and are called pseudo-inverses.[5.3]

We will demonstrate this on the example of the [2 × 2] τ-matrices introduced above (5.4). These matrices differ from the [2 × 2] Pauli matrices in that the numbers *1* are replaced by the minor matrices $\mathbf{I}_R$ and $\mathbf{I}_L$ in the upper row. *In matrix algebraic terms* we should take into account the singularity of these minor matrices (and extend it to [4 × 4] matrices). *In group-theoretical terms,* it is not required that the constituents in the matrices, forming elements of a group, should appear only as numbers. Therefore, when investigating the group properties, we will not consider the special properties of $\mathbf{I}_R$ and $\mathbf{I}_L$.

Theorem 2: τ are self-adjoint matrices.

Definitions: Adjoint matrices are defined in two different ways. *In a strict mathematical sense*, adjoint matrices to $\mathbf{A}_{ij}$ (= AdjA_{ij}) are transposed matrices with

5.3 See more on pseudo-inverses also at Meister et al. (2014), Mostafasadeh (2002–2014), and Ahmed et al. (2003).

signature $(-1)^{i+j}$ composed of the determinants of the minor matrices A_{ji} belonging to the elements a_{ij} of $\mathbf{A}_{ij}$. This interpretation is used for the definition of the inverse matrix of $\mathbf{A}_{ij}$. *In physics*, Hermitic matrices are interpreted as adjoints to $\mathbf{A}_{ij}$ in the sense of transposed matrices composed of the complex conjugate elements $\overline{a_{ji}}$ of $\mathbf{A}_{ij}$ (and are denoted by $\mathbf{A}^{+}$). This interpretation is used in the course of the transformation of scalar vector products, and for defining *self-adjoint* physical operators. The two interpretations lead to different adjoint matrices.

Proof: In the first sense, there holds for 2 × 2 matrices:

$$\mathrm{Adj}\mathbf{M}_{2x2} = -\mathbf{M} + \mathrm{Tr}(\mathbf{M}) \cdot [\mathbf{1}],$$

where Adj**M** denotes the mathematical adjoint of **M**; Tr(**M**) denotes the trace of the matrix; and [**1**] denotes the identity matrix.

$\mathrm{Tr}(\tau_1)$ and $\mathrm{Tr}(\tau_2) = 0, \mathrm{Tr}(\tau_3) = \mathrm{Tr}(\mathbf{I}_\mathrm{L}) - 1 = 0$, which means

$$\mathrm{Adj}\tau_i = -\tau_i \ (i = 1, 2, 3) \tag{5.6}$$

This coincides with the properties of Pauli's σ matrices. (Note that here we did not consider the singularity of $\mathbf{I}_L$ and $\mathbf{I}_R$ since the applied definition for adjunction holds for all 2 × 2 matrices independent of the singularity of minor matrices appearing in their elements.)

In the Hermitic adjoint sense $\sigma_i^{+} = \sigma_i$ and $\tau_i^{+} = \tau_i$, that means, both are self-adjoint in physical terms.

□

Theorem 3: The determinants of the τ-matrices coincide with those of Pauli's σ matrices (–1).

Proof: Instead of reading the determinants from the form of the matrices in (5.4) (which would be $-\mathbf{I}_R$, $-\mathbf{I}_R$, $-\mathbf{I}_L$, respectively, that are not numbers, what would be expected from determinants), let us apply the following definition: $\mathbf{M}\cdot\mathrm{Adj}\mathbf{M} = \det\mathbf{M}\cdot\mathbf{E}$, where **E** is the unit matrix of the τ-algebra.
Considering also (5.6):

$$\tau_i \cdot \mathrm{Adj}\tau_i = -\tau_i^2 = -\mathbf{E} \quad \text{(here } i \text{ is not a summing index)}, \tag{5.7}$$

and comparing this with the definition, one can read from here that

$$\det\tau_i = -1 \ (i = 1, 2, 3), \tag{5.8}$$

this property coincides again with that of the Pauli matrices.

□

Theorem 4: The τ-matrices are unitary.

Proof: Similar to the Pauli matrices ($\sigma_i^+\sigma_i = \sigma_i\sigma_i^+ = [\mathbf{1}]$) the [2 × 2] tau matrices fulfil the (Hermitic sense) unitarity condition: $\tau_i^+\tau_i = \tau_i\tau_i^+ = \mathbf{E}$, where **E,** in terms defined in Eq. (5.4), represents the unit matrix of the τ-algebra. Since, similar to τ_i, the [4 × 4] form of **E** is singular, the group composed by the tau matrices will be called pseudo-unitary.

□

Theorem 5: The τ-matrices are the inverses of themselves.

Proof: In order to check the theorem, read (cf., proof of *Theorem 1*) (pseudo-)inverse of the τ_i matrices:

$$\tau_i^{-1} = \frac{\mathrm{Adj}\tau_i}{\det \tau_i} = \frac{-\tau_i}{-1} = \tau_i \tag{5.9}$$

□

5.2.4 Representation of the group composed by the τ-matrices

Similar to the Pauli matrices, let us compose a representation of the τ-algebra in the following way:

$$\mathbf{K}_+ = (\mathbf{K}_1 + i\mathbf{K}_2);\ \mathbf{K}_- = (\mathbf{K}_1 - i\mathbf{K}_2);\ \mathbf{K}_3 = \frac{1}{2}\tau_3 \tag{5.10}$$

where $\mathbf{K}_i = \frac{1}{2}\tau_i$;

so

$$\mathbf{K}_+ = \begin{bmatrix} 0 & 0 & 0 & 1 \\ 0 & 0 & 0 & 1 \\ 0 & 0 & 0 & 1 \\ 0 & 0 & 0 & 0 \end{bmatrix} = \begin{bmatrix} 0 & \mathbf{I}_R \\ 0 & 0 \end{bmatrix};\ \mathbf{K}_- = \begin{bmatrix} 0 & 0 & 0 & 0 \\ 0 & 0 & 0 & 0 \\ 0 & 0 & 0 & 0 \\ 1 & 0 & 0 & 0 \end{bmatrix} = \begin{bmatrix} 0 & 0 \\ 1 & 0 \end{bmatrix}; \tag{5.11}$$

$$\mathbf{K}_3 = \begin{bmatrix} 1/2 & 0 & 0 & 0 \\ 1/2 & 0 & 0 & 0 \\ 1/2 & 0 & 0 & 0 \\ 0 & 0 & 0 & -1/2 \end{bmatrix} = \begin{bmatrix} \mathbf{I}_L/2 & 0 \\ 0 & -1/2 \end{bmatrix}$$

where

$$[\mathbf{K}_3,\mathbf{K}_+]=\mathbf{K}_+;\quad [\mathbf{K}_3,\mathbf{K}_-]=-\mathbf{K}_-;\quad [\mathbf{K}_+,\mathbf{K}_-]=2\mathbf{K}_3;$$

$$\mathbf{K}_+\mathbf{K}_-=\begin{bmatrix}1&0&0&0\\1&0&0&0\\1&0&0&0\\0&0&0&0\end{bmatrix}=\begin{bmatrix}\mathbf{I}_L&0\\0&0\end{bmatrix};\quad \mathbf{K}_-\mathbf{K}_+=\begin{bmatrix}0&0&0&0\\0&0&0&0\\0&0&0&0\\0&0&0&1\end{bmatrix}=\begin{bmatrix}0&0\\0&1\end{bmatrix} \tag{5.12}$$

$$\mathbf{K}_++\mathbf{K}_-=\tau_1;\quad \mathbf{K}_+-\mathbf{K}_-=i\tau_2;\quad \mathbf{K}_+\mathbf{K}_--\mathbf{K}_-\mathbf{K}_+=\tau_3;\quad \mathbf{K}_+\mathbf{K}_-+\mathbf{K}_-\mathbf{K}_+=\mathbf{E}$$

Theorem 6: $\mathbf{K}^2$ is Casimir invariant.

Proof: Composing the sum
$\mathbf{K}^2=\mathbf{K}_1^2+\mathbf{K}_2^2+\mathbf{K}_3^2=\frac{1}{2}(\mathbf{K}_+\mathbf{K}_-+\mathbf{K}_-\mathbf{K}_+)+\mathbf{K}_3^2=\frac{1}{4}\left(\tau_1^2+\tau_2^2+\tau_3^2\right)=\frac{3}{4}\mathbf{E}$
we get that $\mathbf{K}^2$ is proportional to the identity map (in our case the identity is represented by the unit matrix $\mathbf{E}$ of the group), so, it is a Casimir invariant that commutes with all the three generators of the representation: $[\mathbf{K}^2,\mathbf{K}_i]=0\ (i=1,2,3)$.

□

Theorem 7: The eigenvalues belonging to the $\mathbf{K}$ operators are half-integer.

Proof: We denote the weight of this representation belonging to the eigenvector χ by $k^{(\chi)}$ and that belonging to the eigenvector ϑ by $k^{(\vartheta)}$:

$$\begin{aligned}\mathbf{K}_3\chi&=\frac{1}{2}\tau_3\chi=\frac{1}{2}\mathbf{I}_L\chi=k^{(\chi)}\chi;\quad k^{(\chi)}=\frac{\mathbf{I}_L}{2}\\ \mathbf{K}_3\vartheta&=\frac{1}{2}\tau_3\vartheta=-\frac{1}{2}\vartheta=k^{(\vartheta)}\vartheta;\quad k^{(\vartheta)}=-\frac{1}{2}\end{aligned} \tag{5.13}$$

□

The effects of the $\mathbf{K}_\pm$ operators on the eigenvectors are as follows:

$$\mathbf{K}_+\chi=0,\quad \mathbf{K}_-\chi=-i\vartheta,\quad \mathbf{K}_+\vartheta=\chi,\quad \mathbf{K}_-\vartheta=0 \tag{5.14}$$

This representation shows similitude to a representation of the Pauli matrices, considering the same restrictions for the $\mathbf{I}_L$ and $\mathbf{I}_R$ minor matrices as we showed.

5.2.5 The group of HySy

All properties of this representation *formally coincide* with a representation of the SU(2) group of the Pauli σ matrices. Similarly, it is a special unitary group with two independent parameters, with the condition that its unitary group element is defined as

$\mathbf{E} = \begin{bmatrix} \mathbf{I}_L & 0 \\ 0 & 1 \end{bmatrix}$ (whose $\mathbf{I}_L$ was given in (5.4) and is a singular minor matrix in the 4 × 4 **E**).
In the latter, restricted sense [considering (5.6), (5.8), and (5.9)] the group formed by the τ-matrices is a special, *pseudo-unitary,* two parametric group. We show that this special, pseudo-unitary group is isomorphic with the usual SU(2) group (e.g. among others of the Pauli matrices). The τ-matrices behave like spinor elements of an SU(2) group.

The peculiarities of the τ-matrices (that distinguish them from the Pauli matrices) allow them to make a *correspondence between scalars* and *vector components.*

Theorem 8: The group of the τ-matrices is *pseudo-unitary* and is *isomorphic* with the special unitary SU(2) group.

Definition (a): An operator τ in a space V is defined pseudo-unitary if it preserves the form of the scalar product of two vectors: $<v\tau, w\tau> = <v, w>$, where $\tau \in U(V)$. The pseudo-unitary group $U(n\text{-}l, l)$ or $U(V)$ is the group of all operators satisfying this condition. If the scalar product is positive definite (i.e. $l = 0$), then $U(n,0) = U(n)$ is the normal unitary group (Neretin, 2011, chapter 2) (Onishchik, Sulanke, 2006, chapter 2.1.5).

Definition (b): For τ is an invertible operator, one can easily check a more simple condition, namely that all τ_i and **E** satisfy that the scalar products be $<v\tau, v\tau> = <v, v>$ for all v (provided that the vectors v are expressed on an orthogonal four-dimensional basis when all products $v_i v_j$ vanish if $i \neq j$). In this case, τ is pseudo-unitary (Neretin, 2011, Proposition 2.1, p.68). In short, according to this condition, the τ operators *can* form a (special) pseudo-unitary group.

Proof 1: In the case of the group of our τ operators, $n = 2$, $l = 0$ or 1. The corresponding groups are either $U(2, 0) = U(2)$ or $U(1,1)$.[5.4] The latter option may arise in case of a τ_3 representation (whose main diagonal contains two opposite signed elements, i.e. $l = 1$). $U(1,1)$ needs to use an infinite-dimensional Hilbert space because being a non-compact Lie group, $U(1,1)$ does not admit a finite-dimensional unitary representation (Mostafazadeh, 2004). Consequently, the pseudo-unitary group of the τ operators is isomorphic with the unitary $U(2)$ group. As it was shown earlier, one can assign a |1| absolute value determinant to the (otherwise singular) τ-matrices. In this sense, the τ operators form a group that is isomorphic with the special unitary SU(2) group. We denote it so in the following.

□

5.4 Ahmed and Jain (2003b) showed that a class of group of pseudo-unitary operators is isomorphic to one of the groups $U(n)$ or $U(n,m)$ for some $m,n \in Z^+$.

Proof 2: A matrix in SU(2) must have the form

$$M=\begin{bmatrix} z & w \\ -\bar{w} & \bar{z} \end{bmatrix}$$

where z and w are complex and $\bar{z}$ denotes the complex conjugate of z. To be in SU(2) it is required that $M^{+}= M^{-1}$ and that Det(M) is unitary, where M^{+} is the conjugate transpose of M, and Det denotes determinant. Thus, if $z= t_4+ it_3$ and $w= t_2+ it_1$ where t_4, t_3, t_2, t_1 are real, and $i^2= -1$, then

$$M=\begin{bmatrix} t_4+it_3 & t_2+it_1 \\ -t_2+it_1 & t_4-it_3 \end{bmatrix}$$

with the condition $t_4^{\,2} + t_1^{\,2} + t_2^{\,2} + t_3^{\,2}= 1$.

In our case we can write:

$$M=\begin{bmatrix} (t_4+it_3)\mathbf{I}_L & (t_2+it_1)\mathbf{I}_R \\ (-t_2+it_1)\mathbf{1} & (t_4-it_3)\mathbf{1} \end{bmatrix}=$$

$$=t_1\begin{bmatrix} & i\mathbf{I}_R \\ i & \end{bmatrix}+t_2\begin{bmatrix} & \mathbf{I}_R \\ -1 & \end{bmatrix}+t_3\begin{bmatrix} i\mathbf{I}_L & \\ & -i \end{bmatrix}+t_4\begin{bmatrix} \mathbf{I}_L & \\ & 1 \end{bmatrix}$$

$$M=t_4\mathbf{E}+it_1\tau_1+it_2\tau_2+it_3\tau_3$$

where $\mathbf{I}_L$ and $\mathbf{I}_R$ are [3 × 3] chiral unit matrices, and $\mathbf{1}$ is [1 × 1] unit matrix.

In order to meet the condition $t_4^{\,2}+t_1^{\,2}+t_2^{\,2}+t_3^{\,2}= 1$, let us choose $t_1=t_2=t_3=-1/2,\ t_4=1/2$.

Let us introduce the notation $-i\tau_i= \tau_i^{\,q}$, where $\tau_i^{\,q}$ (i= 1, 2, 3) follow quaternion algebra. This means $(\tau_i^{\,q})^2=-\mathbf{E}$, $\tau_i^{\,q}\,\tau_j^{\,q}= \tau_k^{\,q}$ and $\tau_j^{\,q}\,\tau_i^{\,q}=-\tau_j^{\,q}\,\tau_i^{\,q}$ (i, j, k are cyclic indices), $\tau_i^q\tau_j^q\tau_k^q=-\mathbf{E}$.

Then

$$M=\frac{1}{2}[\mathbf{E}-i\tau_1-i\tau_2-i\tau_3]=\frac{1}{2}\left\{\mathbf{E}-\begin{bmatrix} & i\mathbf{I}_R \\ i & \end{bmatrix}-\begin{bmatrix} & \mathbf{I}_R \\ -1 & \end{bmatrix}-\begin{bmatrix} i\mathbf{I}_L & \\ & -i \end{bmatrix}\right\}=$$

$$=\frac{1}{2}\begin{bmatrix} (1+i)\mathbf{I}_L & (1+i)\mathbf{I}_R \\ (-1+i)\mathbf{1} & (1-i)\mathbf{1} \end{bmatrix}=\frac{1}{2}\left\{\mathbf{E}+\begin{bmatrix} & -i\mathbf{I}_R \\ -i & \end{bmatrix}+\begin{bmatrix} & -\mathbf{I}_R \\ 1 & \end{bmatrix}+\begin{bmatrix} -i\mathbf{I}_L & \\ & i \end{bmatrix}\right\}$$

and:

$$M=\frac{1}{2}\left[\mathbf{E}+\tau_1^q+\tau_2^q+\tau_3^q\right]$$

For **E**, τ_i^q (i= 1, 2, 3) are quaternions, they follow an algebra that excludes complex products. [For the discussion of the comparison of matrix-, quaternion-, and mixed algebras, see Section 5.1.1 and (Darvas, 2015b).]

(a) The unitarity of the determinant of M is interpreted in the following form:

$$\mathrm{Det}(M) = \frac{1}{2}\begin{vmatrix} (1+i)\mathbf{I}_L & (1+i)\mathbf{I}_R \\ -(1-i)\mathbf{1} & (1-i)\mathbf{1} \end{vmatrix} = 1\cdot(\mathbf{I}_L + \mathbf{I}_R)$$

This is a 1 times product of a special symmetric, unitary formula for the sum of otherwise chiral $\mathbf{I}_L$ and $\mathbf{I}_R$ matrices. $1\cdot(\mathbf{I}_L+\mathbf{I}_R) = 1\cdot\begin{bmatrix} 1 & 0 & 1 \\ 1 & 0 & 1 \\ 1 & 0 & 1 \end{bmatrix}$, where the round bracket ( ) contains a 3 × 3 special symmetric matrix "multiplied" by "1". This unusual unitary form is a condition of the SU(2) compatibility of M.

(b) Furthermore, $M^{-1} = \frac{1}{2}\begin{bmatrix} (i-1)\mathbf{I}_L & -(i+1)\mathbf{I}_R \\ (i-1)\mathbf{1} & (i+1)\mathbf{1} \end{bmatrix}$, because

$$M\cdot M^{-1} = \frac{1}{2}\begin{bmatrix} (1+i)\mathbf{I}_L & (1+i)\mathbf{I}_R \\ -(1-i)\mathbf{1} & (1-i)\mathbf{1} \end{bmatrix}\cdot\frac{1}{2}\begin{bmatrix} (i-1)\mathbf{I}_L & -(i+1)\mathbf{I}_R \\ (i-1)\mathbf{1} & (i+1)\mathbf{1} \end{bmatrix} =$$

$$= \begin{bmatrix} 1 & 0 & 0 & 0 \\ 1 & 0 & 0 & 0 \\ 1 & 0 & 0 & 0 \\ 0 & 0 & 0 & 1 \end{bmatrix} = \mathbf{E}$$

(c) Adj(M) = M^{-1}·Det(M). Since $\mathbf{I}_L^2 = \mathbf{I}_L$, $\mathbf{I}_R^2 = \mathbf{I}_R$, $\mathbf{I}_L\mathbf{I}_R = \mathbf{I}_R$ and $\mathbf{I}_R\mathbf{I}_L = \mathbf{I}_L$, ($\mathbf{I}_L$ and $\mathbf{I}_R$ are projector matrices) there is easy to admit that

$$\mathrm{Adj}(M) = M^{-1}\mathrm{Det}(M) = \frac{1}{2}\begin{bmatrix} (i-1)\mathbf{I}_L & -(i+1)\mathbf{I}_R \\ (i-1)\mathbf{1} & (i+1)\mathbf{1} \end{bmatrix}(\mathbf{I}_L + \mathbf{I}_R)$$

(d) The conjugate transpose of M is $M^{+} = \frac{1}{2}\begin{bmatrix} (i-1)(-\mathbf{I}_L) & -(i+1)\mathbf{1} \\ (i-1)(-\mathbf{I}_R) & (i+1)\mathbf{1} \end{bmatrix}$.

In the forms of both M^{-1} (in (b)) and M^{+} (in (d)), the coefficients of (i–1) and (i+1) are not numbers. Considering that the matrices ± $\mathbf{I}_L$, ± $\mathbf{I}_R$, and **1** are unitary in their own category, M^{-1} and M^{+} formally coincide (not less than τ and σ_i do).

□

In summary, $\{\mathbf{E}, \tau_i^q\}(i=1,2,3)$ generate a group, whose elements are $\{\pm\mathbf{E}, \pm\tau_i^q\}$. Under the discussed conditions, these unit quaternions are identified with a representation of an SU(2) compatible quaternion group. Accordingly, $\{\mathbf{E}, \tau_i\}(i=1,2,3)$ generate also an SU(2) compatible group, a set of elements of whose representation are $\{\pm \mathbf{E}, \pm \tau_i, \pm i\mathbf{E}, \pm i\tau_i\}$. In both cases, the generated groups are pseudo-unitary, and two of their elements are independent.

5.3 Comparing the τ-algebra and the Dirac algebra

5.3.1 The τ and the Dirac (γ) matrices

Dirac (1928) assigned a set of [4 × 4] ρ_i matrices to the [4 × 4] bispinor σ_i matrices by interchanging the second and third rows and the second and third columns. His γ algebra (called generally the Dirac algebra) was defined by the $\gamma_i = \rho_2\sigma_i$ matrices. (Note that there appeared also other representations of the Dirac algebra later.)

Notice that the second and third rows and columns of our τ-matrices coincide! Therefore, if we want to assign a set of matrices to the τ_i matrices – using the analogy of Dirac's ρ_i matrices – they will coincide with the τ_i matrices themselves. Instead of a ρ-σ pair, we have got a τ–τ pair.

1) The structure of the expected eigenvectors (χ and ϑ) determine the structure of the τ-matrices.
2) The analogy to the ρ-σ matrix set doublet demands that the change of the second and third rows and columns produce unchanged τ-matrices to achieve identical matrices and avoid duplication, unlike the ρ-σ pair.
3) The rank 2 of the matrices in the τ-algebra is not surprising. These matrices make a correspondence between TWO physical quantities. One of these quantities is a *one-component* scalar, and the other is a vector – constructed by *three components*. This is reflected in the construction of the τ-matrices.
4) We remind again that τ is the next letter in the Greek alphabet following ρ and σ – this justifies the name of the τ-matrices.

Consequently, if we define – in a similar way ($\gamma_i = \rho_2\sigma_i$) like we obtained the γ matrices – a set of T [read: upper case Greek tau] matrices, we get – by definition – the following: $T_i = \tau_2\tau_i$. Since the τ-matrices transform into each other, we get the following algebra:

$$\begin{aligned} T_1 &= \tau_2\tau_1 = -i\tau_3 \\ T_2 &= \tau_2\tau_2 = \mathbf{E} \qquad \text{(left – handed “unit” operator)} \\ T_3 &= \tau_2\tau_3 = i\tau_1 \end{aligned} \tag{5.15}$$

Let us define $T_4 = \tau_4^2 = \tau_3$, where $\tau_4 = \begin{bmatrix} 1 & 0 & 0 & 0 \\ 1 & 0 & 0 & 0 \\ 1 & 0 & 0 & 0 \\ 0 & 0 & 0 & i \end{bmatrix}$

This set defined another representation of the *tau* algebra that we denote by the upper case Greek *T*. The *T*-algebra should behave similarly over a field (that we will apply later in this book to describe a property of the field of the IFC) like the γ algebra did over the bispinors' space.

Applying the (5.15) defined T-matrices expressed with τ's, we get the following algebra for the transformation of the eigenvectors χ and ϑ:

$$\begin{aligned} T_1\chi &= -i\chi & T_1\vartheta &= i\vartheta \\ T_2\chi &= \chi & T_2\vartheta &= \vartheta \\ T_3\chi &= \vartheta & T_3\vartheta &= -\chi \\ T_4\chi &= \chi & T_4\vartheta &= -\vartheta \end{aligned} \tag{5.16}$$

5.3.2 Comparison of the algebras of the δ- and the τ-matrices

Let us introduce the following notations:

$$\delta_1 = \tau_1^2; \quad \delta_2 = \tau_2^2; \quad \delta_3 = \tau_3^2; \quad \delta_4 = \tau_4^2.$$

Similar to the **K** representation of the τ-matrices, let us compose an **L** representation of the δ-matrices.

The effects of the $\delta_1 = \delta_2 = \delta_3 = \delta_i = \tau_i^2 = \mathbf{E}$ and the $\delta_4 = \tau_3$ matrices on the χ and ϑ eigenvectors are as follows:

$$\begin{aligned} \delta_i\chi &= \chi & \delta_i\vartheta &= \vartheta \\ \delta_4\chi &= \chi & \delta_4\vartheta &= -\vartheta \end{aligned}$$

Define

$$\mathbf{L}_+ = \frac{1}{2}(\delta_i + \delta_4) = \begin{bmatrix} I_L & 0 \\ 0 & 0 \end{bmatrix} = \mathbf{K}_+\mathbf{K}_-$$

$$\mathbf{L}_- = \frac{1}{2}(\delta_i - \delta_4) = \begin{bmatrix} 0 & 0 \\ 0 & 1 \end{bmatrix} = \mathbf{K}_-\mathbf{K}_+$$

then

$\mathbf{L}_+ + \mathbf{L}_- = \mathbf{E} = \delta_i$ and $\mathbf{L}_+ - \mathbf{L}_- = \tau_3 = \delta_4$ (where i = 1, 2, 3)

The $\mathbf{L}_+$ and $\mathbf{L}_-$ operators are represented by projector matrices $\mathbf{L}_+^2 = \mathbf{L}_+; \mathbf{L}_-^2 = \mathbf{L}_-$. Furthermore, in contrast to the **K** operators, they are orthogonal: $\mathbf{L}_+\mathbf{L}_- = \mathbf{L}_-\mathbf{L}_+ = 0$.

Note, that the $\mathbf{L}_+$ and $\mathbf{L}_-$ operators are expressed in terms of a τ_3 representation of the tau algebra (by the help of its unit matrix **E** and the τ_3 matrix). The δ_i (=**E**) and δ_4 (=τ_3) matrices compose a group, in contrast to the γ matrices of the Dirac algebra that do not.

The effects of the $\mathbf{L}_+$ and $\mathbf{L}_-$ matrices on the χ and ϑ eigenvectors are as follows:

$$\mathbf{L}_+\chi = \chi \quad \mathbf{L}_+\vartheta = 0$$
$$\mathbf{L}_-\chi = 0 \quad \mathbf{L}_-\vartheta = \vartheta$$

that means, $\mathbf{L}_+$ acts only on χ and the orthogonal $\mathbf{L}_-$ acts only on ϑ. Their effect is the opposite of the effects of the $\mathbf{K}_\pm$ operators. However, the effects of the δ-matrices and the $\mathbf{L}$ matrix representation do not modify the $\pm 1/2$ eigenvalues of the χ and ϑ eigenvectors revealed in Eq. (5.13).

The effect of the two types of δ-matrices on the eigenvectors χ and ϑ of the operators of the $\boldsymbol{\tau}$-algebra can be expressed in the following short [8 × 8] matrix form, where δ_i and δ_4 are [4 × 4] matrices that act on the four-element vector columns of χ and ϑ:

$$\begin{bmatrix} \delta_i & 0 \\ 0 & \delta_4 \end{bmatrix} \begin{bmatrix} \chi & \vartheta \\ \chi & \vartheta \end{bmatrix} = \begin{bmatrix} \chi & \vartheta \\ \chi & -\vartheta \end{bmatrix} \tag{5.17}$$

The δ_i matrices leave intact the eigenvectors χ and ϑ, while δ_4 changes the sign of one of the eigenvectors, ϑ.

5.3.3 The τ algebra beyond physics (matrix genetics)

We show an example that demonstrates the applicability of the τ algebra beyond physics. It has been applied successfully in matrix genetics. Genetic algebra shows 2+2 type character defined in a purine–pyrimidine field, and 1+3 type character defined in an RNA–DNA field. Therefore, similar matrices can demonstrate the 1+3 character in genetic algebra like those that govern the matrices elaborated for QED. The algebraic methods and transformation formula developed in the theory of physical fields, like in this section, appear also in the transformation properties of certain genetic matrices.

S. Petoukhov (2011, 2012, 2015a) referred several times to analogies between the algebra of the genetic code and the algebra applied in QED. The latter matrix transformation confirms those findings. To understand it, let us try to replace the two χ four-columns in the left of the matrix for the two purine traits (G, A), and the ϑ four-columns in the right of the matrix for the two pyrimidine traits (C, T) in the DNA bonds in the Eq. (5.17). In terms of Hadamard matrices (Petoukhov, 2011, 2008), this transformation coincides with the transformation of the nitrogenous base uracil (U) into another nitrogenous base thymine (T) when one changes from an RNA to DNA during a recombination process, what strictly corresponds to the sign change of the ϑ in the bottom right position in the 2 × 2 matrix in the right side of the Eq. (5.17).

As Petoukhov (2012a, p. 389) writes: An “interesting structural feature of the 8 octets of triplets is connected with the phenomenon of the special status of the T (thymine) in the basic alphabet of DNA. Among the four DNA bases – A, C, G, T – the letter T contrasts phenomenologically with three other letters of the alphabet: 1) only the letter T is transformed into another letter U (uracil) in the transition from DNA to RNA; 2) only the letter T (and its changer U) has not the functionally important amino group NH_2 in contrast of other three letters . . . . This binary opposition can be expressed in a digital form." Similar to our separation of the roles of field-charges and velocities in the charge currents, the DNA bases can be associated as A, C, and G with a sign +1 and T with a sign −1. Then "each triplet under replacing its letters on these numbers (A = C = G = +1, T = −1) can be represented as the product of these numbers. For example, the triplet CAT is represented as 1*1*(−1) = − 1 and the triplet TGT – as (−1)*1*(−1) = +1. In the result, the 8 octets of triplets obtain numerical representations as sequences of elements +1 and − 1. . . . The set of these sequences coincide with the complete system of orthogonal Walsh functions for 8-dimensional spaces".

The golden matrices by S. Petoukhov are derived from his matrices **P** that include the number of hydrogenous bonds coupling the nucleotides (and other proportional properties of the nucleotides) (cf., Petoukhov, 2006, 2008). The basic matrix is

$$\mathbf{P}_{\mathbf{MULT}}{}^{(\mathbf{1})} = \begin{vmatrix} 3 & 2 \\ 2 & 3 \end{vmatrix}$$

and its tensorial powers. The golden matrices are defined as the matrix square roots of the consecutive power genetic matrices $\mathbf{P}_{\mathrm{MULT}}{}^{(n)}$. (On the DNA sequence numerical representation, see also Abo-Zahhad et al. (2014).) It is easy to check that all elements of these square root matrices are expressed in powers of the golden ratio $\varphi = (1+\sqrt{5})/2$. For example:

$$(\mathbf{P}_{\mathrm{MULT}})^{1/2} = = \mathbf{\Phi}_{\mathrm{MULT}} = \begin{vmatrix} \varphi & \varphi^{-1} \\ \varphi^{-1} & \varphi \end{vmatrix}; \quad \left(\mathbf{P}_{\mathrm{MULT}}{}^{(2)}\right)^{1/2} = \\ = \mathbf{\Phi}_{\mathrm{MULT}}{}^{(2)} = \begin{vmatrix} \varphi^2 & \varphi^0 & \varphi^0 & \varphi^{-2} \\ \varphi^0 & \varphi^2 & \varphi^{-2} & \varphi^0 \\ \varphi^0 & \varphi^{-2} & \varphi^2 & \varphi^0 \\ \varphi^{-2} & \varphi^0 & \varphi^0 & \varphi^2 \end{vmatrix}$$

and so on.

The matrix $\Phi^{(2)}_{\mathrm{MULT}}$ can be decomposed:

$$\Phi^{(2)}_{\mathrm{MULT}} = \begin{bmatrix} \varphi^2 & \varphi^0 & \varphi^0 & \varphi^{-2} \\ \varphi^0 & \varphi^2 & \varphi^{-2} & \varphi^0 \\ \varphi^0 & \varphi^{-2} & \varphi^2 & \varphi^0 \\ \varphi^{-2} & \varphi^0 & \varphi^0 & \varphi^2 \end{bmatrix} =$$

$$= \varphi^2 \begin{bmatrix} 1 & 0 & 0 & 0 \\ 0 & 1 & 0 & 0 \\ 0 & 0 & 1 & 0 \\ 0 & 0 & 0 & 1 \end{bmatrix} + \varphi^0 \begin{bmatrix} 0 & 0 & 1 & 0 \\ 0 & 0 & 0 & 1 \\ 1 & 0 & 0 & 0 \\ 0 & 1 & 0 & 0 \end{bmatrix} + \varphi^{-2} \begin{bmatrix} 0 & 0 & 0 & 1 \\ 0 & 0 & 1 & 0 \\ 0 & 1 & 0 & 0 \\ 1 & 0 & 0 & 0 \end{bmatrix} + \varphi^0 \begin{bmatrix} 0 & 1 & 0 & 0 \\ 1 & 0 & 0 & 0 \\ 0 & 0 & 0 & 1 \\ 0 & 0 & 1 & 0 \end{bmatrix}.$$

The matrix $\Phi^{(2)}_{MULT}$ is weighted (with the powers of φ), so they are not unitary. (On the golden matrices, see also Wani and Badshah (2017).) In their representation, the respective vectors in a field do not conserve their length under transformation. Consequently, they will not subject to any form of Lorentz transformation. At the same time, they show considerable symmetries and similarities in their transformations. The most important is, probably, that they can be composed of a part of the elements of the same spinor matrices which we learned in the QED (cf., Section 5.1.1). Moreover, they are all real and include fewer spinor elements than the genetic matrices do: this is the price that we earned in response to the loss of the unitarity (see also in Darvas and Petoukhov, 2019).

At a first approximation, the matrix $\Phi^{(2)}_{MULT}$ can be constructed by the help of the [2 × 2] Pauli σ_0 and σ_1 matrices.

$$\Phi^{(2)}_{MULT} = \varphi^2 \begin{bmatrix} \sigma_0 & 0 \\ 0 & \sigma_0 \end{bmatrix} + \varphi^0 \begin{bmatrix} 0 & \sigma_0 \\ \sigma_0 & 0 \end{bmatrix} + \varphi^{-2} \begin{bmatrix} 0 & \sigma_1 \\ \sigma_1 & 0 \end{bmatrix} + \varphi^0 \begin{bmatrix} \sigma_1 & 0 \\ 0 & \sigma_1 \end{bmatrix}.$$

The symmetry is obvious. However, the two matrices in the middle are not members of Dirac's [4 × 4] σ matrices. They are defined by the help of Dirac's matrix ρ_1 (see also Darvas, 2015b, 2018b). It is the matrix that may change the values of the first and second columns of the matrices for the third and fourth columns (and *vice versa*), what makes correspondence between Petoukhov's U-complex numbers (Petoukhov, 2017, 2018) and marks a strong relation to the genetic matrices (Petoukhov, 2006, 2008):

$$\begin{bmatrix} \sigma_0 & 0 \\ 0 & \sigma_0 \end{bmatrix} = \sigma_0; \begin{bmatrix} 0 & \sigma_0 \\ \sigma_0 & 0 \end{bmatrix} = \rho_1; \begin{bmatrix} 0 & \sigma_1 \\ \sigma_1 & 0 \end{bmatrix} = \rho_1 \sigma_1; \begin{bmatrix} \sigma_1 & 0 \\ 0 & \sigma_1 \end{bmatrix} = \sigma_1$$

One can recognise in the form of the matrix $\Phi^{(2)}_{MULT}$ again that it will not transform according to the Lorentz transformation. However, the just derived symmetric form with the help of the σ and ρ_1 matrices (ρ_1 formed by Dirac from σ_1) deserves interest to present the transformation rule of $\Phi^{(2)}_{MULT}$.

As mentioned earlier, the genetic matrices appear simultaneously in the (2+2) dimensional purine–pyrimidine field, and the (1+3) dimensional RNA–DNA field. The latter means that while the uracil (*U*) transforms into thymine (*T*) during the transformation from an RNA to DNA, the other three letters of the genetic alphabet do not. In order to describe the combined transformation in the two genetic fields, one should define the matrices of the combined transformation.

First, let us define an abstract field over the Minkowski space, with vectors $\boldsymbol{z}$ in it. Transformations of the vectors $\boldsymbol{z}$ can be composed by the following similar (but not Lorentzian) transformation $F(z^\mu)$:

$$F(z^\mu) = z^\mu \zeta_\mu$$

$$F^{(1)}(z) = z^i \begin{bmatrix} \varphi & \varphi^{-1} \\ \varphi^{-1} & \varphi \end{bmatrix} = z^0\varphi[\sigma_0] + z^1\varphi^{-1}[\sigma_1]$$

$$F^{(2)}(z^\mu) = z^0\varphi^2[\sigma_0] + z^1\varphi^0[\rho_1] + z^2\varphi^{-2}[\rho_1\sigma_1] + z^3\varphi^0[\sigma_1] =$$

$$= \begin{bmatrix} \varphi^2 z^0 + z^3 & z^1 + \varphi^{-2}z^2 \\ z^1 + \varphi^{-2}z^2 & \varphi^2 z^0 + z^3 \end{bmatrix}$$

where $\zeta_0 = \varphi^2\sigma_0$; $\zeta_1 = \varphi^0\rho_1$; $\zeta_2 = \varphi^{-2}\rho_1\sigma_1$; $\zeta_3 = \varphi^0\sigma_1$.

It is easy to see that the sign of the z^1 components in the side diagonal of the matrix $F^{(2)}(z^\mu)$ differs from that expected in a Lorentz transformation. However, we have still considered only a transformation in the *purine–pyrimidine field*. To find the proper transformation that may change the sign of z^1 but leaves the sign of the other three genetic letters intact, we should combine it with a transformation in the *RNA–DNA field.* Since it has a (1+3) character, we must turn to the tau (τ) algebra of HySy as it is described in Section 5.2. According to the theory of HySy, applied in (–,+,+,+) signature and in τ_3 representation, two matrices govern the transformation in a (1+3) field:

$$E = \begin{bmatrix} 1 & 0 & 0 & 0 \\ 0 & 0 & 0 & 1 \\ 0 & 0 & 0 & 1 \\ 0 & 0 & 0 & 1 \end{bmatrix} \text{ and } \tau_3 = \begin{bmatrix} -1 & 0 & 0 & 0 \\ 0 & 0 & 0 & 1 \\ 0 & 0 & 0 & 1 \\ 0 & 0 & 0 & 1 \end{bmatrix}.$$

For the combined transformation, we introduce a combined transformation matrix, denoted by ξ. Notice, that similar to the role of σ_0 in the set of the Pauli matrices, E coincides also with the unit matrix in the HySy. They and their combined application leave the subjected matrices intact. Let us concentrate our attention on the role of τ_3. When U is in position 1, τ_3 should be combined with σ_1. (When a permutation in the matrix $\begin{bmatrix} C & U \\ A & G \end{bmatrix}$ places U to another position, σ_1 should be replaced by ρ_1 and $\rho_1\sigma_1$, respectively.) So the matrix of the combined transformation will be the following:

$$\xi = \sigma_1\tau_3 = \begin{bmatrix} 0 & 1 & 0 & \\ 1 & 0 & 0 & 0 \\ 0 & 0 & 0 & 1 \\ 0 & 0 & 1 & 0 \end{bmatrix} \begin{bmatrix} -1 & 0 & 0 & 0 \\ 0 & 0 & 0 & 1 \\ 0 & 0 & 0 & 1 \\ 0 & 0 & 0 & 1 \end{bmatrix} = \begin{bmatrix} 0 & 0 & 0 & 1 \\ -1 & 0 & 0 & 0 \\ 0 & 0 & 0 & 1 \\ 0 & 0 & 0 & 1 \end{bmatrix}.$$

The application of this transformation allows to change the sign of U and, in the transformation of $F^{(2)}(z^\mu)$, the sign of z^1. Now, the following transformations should be applied instead of those that appeared in ζ_μ earlier:

$$\zeta'_0 = \varphi^2\xi\sigma_0; \quad \zeta'_1 = \varphi^0\xi\rho_1; \quad \zeta'_2 = \varphi^{-2}\xi\rho_1\sigma_1; \quad \zeta'_3 = \varphi^0\xi\sigma_1$$

where

$$\xi\sigma_0 = \begin{bmatrix} 0 & 0 & 0 & 1 \\ -1 & 0 & 0 & 0 \\ 0 & 0 & 0 & 1 \\ 0 & 0 & 0 & 1 \end{bmatrix}; \; \xi\rho_1 = \begin{bmatrix} 0 & 1 & 0 & 0 \\ 0 & 0 & -1 & 0 \\ 0 & 1 & 0 & 0 \\ 0 & 1 & 0 & 0 \end{bmatrix};$$

$$\xi\rho_1\sigma_1 = \begin{bmatrix} 1 & 0 & 0 & 0 \\ 0 & 0 & 0 & -1 \\ 1 & 0 & 0 & 0 \\ 1 & 0 & 0 & 0 \end{bmatrix}; \; \xi\sigma_1 = \begin{bmatrix} 0 & 0 & 1 & 0 \\ 0 & -1 & 0 & 0 \\ 0 & 0 & 1 & 0 \\ 0 & 0 & 1 & 0 \end{bmatrix}.$$

Although in a general case, this is not a length conserving transformation, and also not a proper Lorentz transformation in its classical sense, formally similar, we gained, on the other hand, other advantages:

(a) It shows more (another kind) symmetry and demonstrates the relation both to the QED's algebra (cf., Section 11.3) and the genetic algebra.
The last derived transformation demonstrates that

(*b*) the proper transformation between an RNA and DNA can algebraically be described by a combined transformation applied in the purine–pyrimidine and the RNA–DNA fields simultaneously (similar to the method applied for the extended Dirac equation in Darvas (2013), and in Section 11.3 of this book).

The statements (*a*) and (*b*) are based on analogue algebras that govern physical field theory and the theory of the genetic codes. This pragmatic similitude is a nice example that certain symmetry methods, expressed in an algebraic form in physics, can be productive in apparently "distant" disciplines.

In summary, the combined transformations applied together in the convolution of two fields (2+2 and 1+3 dimensional) may explain the structure change between RNA and DNA (cf., Darvas and Petoukhov, 2019). For this reason, mathematical analogies were borrowed from HySy applied to QED field theory. Their study demonstrated that there appear some general regularities in nature that present themselves in different domains of our knowledge about nature (here concretely, genetics and physical field theory), which were considered distant and not overlapping in their methods and laws until now. The functioning of that analogy is a sign of general applicability of the τ- and δ-algebras.

5.4 Summary of the τ algebra

We have identified an algebra that may transform 3 and 1 component quantities into each other. They may be applied to transform isotopic states mutually into each other. If the two isotopic states of ר are denoted by +1/2 and −1/2, they could be identified with the eigenvalues of the τ operators. Since the τ operators compose a group isomorphic with the SU(2) group, one can assume some spin-like property for the ר field-charges. The question is, how?

6 VELOCITY DEPENDENCE IN PHYSICS

We discussed the first condition of *raison d'etre* of a hypersymmetric theory, the necessity to assume isotopic field-charge (IFC) twins in Section 4.

One of the members of these IFC siblings ($ℸ_T$) is associated with a kinetic field.

6.1 Velocity-dependent phenomena

We saw in Section 2.5 that the inertial mass of a body moving in a given reference frame increases depending on its velocity with respect to that frame according to the special theory of relativity. The value of the inertial mass depends on its actual velocity with respect to an inertial frame but *not* on where the particle is located in space. Inertial mass is localised in velocity space. In contrast to this, the particle's ability to *attract* another particle, what depends on its *gravitational* mass, does not change with its velocity. That is constant in all reference frames. The value of the gravitational mass does *not* depend on the particle's velocity relative to an inertial reference frame but it *does* depend on the *location* of that particle, relative to the particle it attracts. Hence, the difference between the inertial and the gravitational mass provides enough reason to make a distinction between them. There is not only the transformation rule that discerns the two kinds of elementary masses from each other but also their localisation. The same holds for a set of elementary masses (and their mixed IFC states) as well.

We know many phenomena in classical physics that depend on velocity in a given reference frame. As examples, first, the kinetic energy is mentioned. Further, in classical electrodynamics, for example, the force caused by the current of moving electric charges (Lorentz charges, q_T) depends on the velocity of the charges compared to the reference frame of the observer [$F\sim (v/c)\times B$]. In relativistic field theories, the best-known example for velocity-dependent phenomena is the covariant effect of the Lorentz transformation [$(x^\mu)' = \Lambda^\mu_\nu(v)x^\nu$ for space–time vectors x^ν and $(F^{\mu\nu})' = \Lambda^\mu_\alpha(v)\Lambda^\nu_\beta(v)F^{\alpha\beta}$ for the electromagnetic field tensor $F^{\alpha\beta}$]. In general, we require that all physical theories be invariant under the Lorentz transformation. Descriptions of the mentioned phenomena handle the space–time co-ordinates as indirect variables. The Lorentz invariance depends only on the velocity difference between the compared systems in given reference frames.

In general, kinetic quantities depend first on velocity in the chosen reference frame, and only indirectly, through $v = v(x_i, t)$ on the space–time variables. As Mills (1989) observed, "Hamilton's principle was first discovered in connection with mechanical systems, where the Lagrangian turns out to be the difference between the kinetic and potential energies, but the principle is easily extended to include

https://doi.org/10.1515/9783110713183-006

velocity-dependent forces of certain types", including, for example, the magnetic force on a moving, electrically charged particle.

Remember, we denoted the *isotopic field-charge* as a property by ד$_V$ and ד$_T$ (Hebrew D, dalet) which can be identified in the case of the gravitational field with the properties of the masses of gravity and inertia, respectively, in the case of the electromagnetic field with the properties of the Coulomb, and current- (or Lorentz) like charges, respectively, and so on. The potential isotope of ד (ד$_V$) is associated directly with space–time co-ordinates. The physical state of the kinetic isotope of ד (ד$_T$) is associated primarily with the components of its velocity (and indirectly with its space–time co-ordinates).

6.2 Velocity-dependent fields

When we try to specify physical phenomena that distinguish kinetic behaviour of objects from their behaviour in a field caused by another, potential source (i.e. ד$_V$), we should attempt to seek for a description in a velocity-dependent field.

The second condition of *raison d'etre* of a hypersymmetric theory is the existence of solely velocity-dependent fields among the kinetic ones. The idea of velocity-dependent fields is not very new in physics, either.

Returning to the cited idea by R. Mills, it is not surprising that phenomena related solely to the kinetic part of the Hamiltonian (T) can be described in a velocity-dependent, that is, kinetic field, where the dependence on the local co-ordinates is indirect. We refer to Norton (2003, cited also in Section 6.3) according to whom the passive general covariance of a theory "has delivered us two fields, $\varphi(x^i)$ and $\varphi'(x^i)$. They are not merely two representations of the same field in different coordinate systems. They are defined in the *same* coordinate system and are mathematically distinct fields, insofar their values at given events will (in general) be different. Active general covariance allows the generation of the field $\varphi'(x^i)$ from $\varphi(x^i)$ by the transformation x^i to x'^i " (p. 122) and the same applies, when we use parameters $\varphi(\dot{x}_\mu(x_\nu))$ (and even in more general cases, pp. 118–120). We will denote a velocity-dependent field by $\mathbf{D}_T = \mathbf{D}[v(x_i, t)]$.[6.1] This does not disclose the possibility of localisation of the theory in space–time, however, it does not ensure it automatically. Local symmetry in a kinetic field means that the objects, fields, or physical laws in question are invariant under a local transformation, namely under a set of the continuously infinite number of separate transformations with an arbitrarily different one at every velocity in the given reference frame.

6.1 The letter **A** has been used for the potentials of the electromagnetic field subject of U(1) symmetry, letter **B** for the weak field, subject of SU(2) symmetry, letter **C** for the strong, colour fields, subject of SU(3) symmetry. Velocity-dependent fields represent a next category, which are subject of an SU(2) isomorphic symmetry group. This justifies to denote them by the next letter of the alphabet, **D**.

I mention again that the necessity of a velocity-dependent field was introduced by Dirac in his (1951a; b) theory. His idea was interpreted in Section 3.5. Dirac defined the velocity field through the electromagnetic four-potential. In the same theory, he considered that the classical theories of the electromagnetic field (including his 1928–29 one) are *approximate* and are valid only if the accelerations of the electrons are small (see also in Section 11.3).

Dirac's approach was confirmative, discussed again by de Haas (2005), cited also in Section 3.5. Jentschura and Adhikari (2018) treated again the role of velocity dependence in gauge fields.

Before discussing field transformations (cf., Section 7), we compare transformations in the configuration space $[(x_\mu, \dot{x}_\mu) \to (x'_\mu, \dot{x}'_\mu)]$ with transformations in a gauge field. A transformation in the configuration space can be generalised to functions in the form $[f(x_\mu, \dot{x}_\mu) \to f(x'_\mu, \dot{x}'_\mu)]$. A transformation in a gauge field would be in general $[\psi(x_\mu, \dot{x}_\mu) \to \psi'(x_\mu, \dot{x}_\mu)]$, which demonstrates the difference. However, Utiyama's results guarantee separability in a gauge field in the form: $[\psi_1(x_\mu)\psi_2(\dot{x}_\mu) \to \psi'_1(x_\mu)\psi'_2(\dot{x}_\mu)]$ instead of a coupling $[\psi_1(x_\mu)\psi_2(\dot{x}_\mu) \to \psi_1(x'_\mu)\psi_2(\dot{x}'_\mu)]$. This means we can separate the space–time dependent and the velocity-dependent fields. Applying the separability, we can introduce the combination of a space–time dependent transformation in the configuration space with a velocity-dependent transformation in a gauge field, where $\dot{x}_\mu = \dot{x}_\mu(x_\nu)$. This combined choice allows us to keep the x_μ-s as independent variables that we will use in the implicit derivations of general Lagrangians. In more general cases, $[(x_\mu, \partial_\nu x_\mu) \to (x'_\mu, \partial_\nu x'_\mu)]$ and, respectively, $[\psi(x_\mu, \partial_\nu x_\mu) \to \psi'(x_\mu, \partial_\nu x_\mu) = \psi'_1(x_\mu)\psi'_2(\partial_\nu x_\mu)]$. The so defined coupling of these configuration space and gauge field transformations will be applied in the IFC transformation in the next section.

6.3 Velocity dependence in the light of conservation laws and symmetries

We referred to the expectation that all new developments in physics should be based on certain symmetry principle, already mentioned in the Introduction (Section 1). The new symmetry principles, like the known ones, should be related to conservation laws (Noether, 1918). However, there appeared always objections in the physical literature on the *role* of symmetries and conservation laws. The question was not about their existence, rather about how they predominate in the discussed theory.

As it is widely known, Wigner (1967) classified invariances into geometrical and dynamical ones. He interpreted, "the primary difference is that the former concern the invariance of *all* the laws of nature under geometric transformations tied to regularities if the underlying spacetime, while the latter concern the form of invariance (i.e. covariance) of the laws governing *particular* interactions under groups of transformations not tied to space-time". (Some divide the dynamical symmetries further into a class with a heuristic potential and another class that are accidental.)

Martin mentions (2003, p. 54) about the ambivalent attitudes toward dynamical invariances: "Perhaps we could take it that the physical significance of the requirement of local gauge invariance lies specifically in that the required gauge potential (s) be dynamical. The common definition of what it is to be dynamical in the context of field theory is for the field in question to appear in the action and for it to be varied in deriving equations of motion, including its own."

According to Weinberg (1996), the real force of the demand of local gauge invariance is carried by the specific requirement that the gauge potential (or the metric connection for gravity) be dynamical in the above sense.

In a late paper, Wigner (1984) questioned whether a gauge invariance is a symmetry principle, meaning that the physical reality of a mathematical gauge invariance may be uncertain. This is a warning that it is not enough to show the existence of a gauge invariance in our physical equations, we must demonstrate the physical relevance of the obtained mathematical result and its consequences. Therefore, we will devote a special subsection (7.5) in this book following the mathematical derivation of the IFC spin gauge invariance to demonstrate its physical relevance.

Adding to Wigner's concern, Martin (2003, p. 55) reminds that in the case of the electromagnetic interaction "the physically loaded step in the argument was the investing of the gauge potential with physical life through the addition to the Lagrangian of a kinetic term describing the free electromagnetic field and the subsequent variation of the gauge potential. . . . this 'dynamization' was in large part affected by hand, rather than necessitated by the local invariance requirement. Thus, any argument to the effect that the *dynamical* gauge field is associated with a, therefore, physical local invariance requirement faces challenges".[6.2,6.3] This challenge is usually tested by experiments that can confirm the reality of the "dynamized" local invariance. We expose our results presented in this work to such tests too. As Martin notices, "there is no universally accepted view concerning what physical content, if any, is to be ascribed to the demand of general covariance" (p. 56) (cf., e.g. Kretschmann's objections to Einstein's notion of covariance).

As regards our relation to the different comprehensions of covariance, we must refer to Norton (e.g. 2003, cf., Section 6.2), who distinguishes weak (or passive)

6.2 Concerns about the observability of local gauge invariances are discussed by Brading and Brown (2004).

6.3 Bohr and Rosenfeld (1950) characterise the dualistic aspect of relastivistically covariant electrodynamics in the following: "[A]n unambiguous definition of the electromagnetic field rests solely on the consideration of the momentum imparted to appropriate test bodies carrying charges or currents, while the charge-current distributions referring to the presence of particles are ultimately defined by the fields to which these distributions give rise." Their approach leads to, what they call as "vanishing interaction between field and particles".

covariance and strong (or active) covariance. Our conservation laws (equations) derived in Section 7.3 will be covariant in the strong (active) sense.

Returning to the earlier concerns, those scruples of Wigner about the physical origin (and thus, even reality) of certain appearances of dynamical invariances were, probably, in concordance with the concerns of Dirac about local gauge invariances and that led the latter to call up the idea of the ether which had been buried with the emergence of the STR many decades earlier. According to the author's opinion, both concerns are connected to the convention on localisation in space–time. In accordance with that convention, both the potential and the kinetic parts of a Hamiltonian (and Lagrangian) must be localised in space–time of the same reference frame. *Why do we insist on this (at least partial) uniformity in the dependence on the same parameters of our physical expressions*? This approach neglected at least two characteristics of the gauge fields and their sources. First, the gauge fields are of *kinetic* character, therefore they are primarily *localised in velocity space* (and indirectly – through the space–time dependence of the velocities – on space–time, determined by the time derivatives and their Lorentz transformation). Second, the assumption of primary localisation of local gauge fields by space–time co-ordinates in a given reference frame contradicts the essence of the local character of a gauge field. The latter means (among others) that there are free parameters which may take different values in the gauge field at the same space–time point in a given reference frame (cf., also Kretschmann's objection to the interpretation of covariance in GTR). Gauge invariances, as shown by E. Noether in her second theorem (and then developed further by Utiyama), emerge due to this difference in the number of the independent and dependent variables.

As Dirac noticed, we introduce extra variables in the theory, and at the same time we bring in a gauge symmetry to extract the physically relevant context. Noether's second theorem says that for a gauge-invariant system – that is, a system invariant under transformations defining a simply connected continuous group $G_{\infty r}$, whose parameters are r arbitrary functions (of another arbitrary number of – and physical meaning, if any – variables), which functions can be derived up to the kth order – there exist r identities of the Euler–Lagrange derivatives of the Lagrange function (Darvas, 2018b). This involves dependencies between the various fields appearing in the theory, that is, there appear less independent equations of motion than the number of unknown parameters (cf., the Bianchi identities), and this is the source of the surplus variables and underdetermination. The circle of problems around the role of those variables whose evolution is unconstrained by the dynamical laws of the given theory is discussed in detail by Belot (2003).

There exists a homomorphy (and not isomorphy) between a gauge field and four-dimensional (4D) space–time, which means, there can be infinitely many points in a gauge field which correspond to a single point in 4D space–time (similar to the case when the points of a higher dimensional space are projected to a lower dimensional one). According to Noether's second theorem, the gauge invariance relations result in this difference in the number of parameters, and further, these invariances are closely related to the "dynamics" of the given invariance. This means, on the one hand, that the observable physical quantities (e.g. field strengths) at a given space–time point do not unambiguously determine a gauge field, or in other words, different values at smoothly changing, infinite number, different points in the gauge field may cause the same event in a given space–time point, and on the other hand, this multiple_causes-single_effect unambiguity does not affect our physical laws based on the observables. The latter guarantees stability in our picture on the physical nature.

Hopefully, the consideration of the difference in the primary parameters on which the potential part of the Hamiltonian (matter field, space–time co-ordinates) and the kinetic part of the Hamiltonian (gauge field, velocity) depend directly – while taking into consideration the *physical* difference between the sources of the two parts, and the description of their transformation into each other (cf., Section 10.1) – eliminates the need for the mentioned concerns.

Nothing is more dangerous than an idea
when it is the only one we have.
Emile Chartier

Chapter III: **ISOTOPIC FIELD-CHARGE SPIN**

7 CONSERVATION LAWS AND HYPERSYMMETRY

7.1 Preliminary assumptions

We would like to justify the existence of distinct isotopic states of field-charges. This assumption requires to conjecture interaction between their isotopic states.

The proposed conceptual framework and assumption on the interaction and its mechanism goes beyond the standard model (SM). Many physicists are convinced that SM does not hold eternally alone and is not untranscendable; there will appear new, more precise theories that will partially include the SM, and answer those questions that are left open by the SM. However, we do not certainly know how, at least at present.

CERN organised three workshops to discuss possible theoretical candidate models beyond the SM to base a "new physics" in accordance with fine-scale anomalies and symmetry breakings in high energy experiments, in 2005–2007 (CERN workshop, 2008a; CERN workshop, 2008b; CERN workshop, 2008c) (and a few others since then). They agreed that SM holds, it needs only some extensions. So do we as well. The present book provides an alternative extension theory, still not discussed in those three working group reports. Nevertheless, much development confirmed the need for a new physics during the past decade.

This work (started in January 2001) is an attempt to exceed a couple of the limits of the SM. As cited in Section 1, Gerard 't Hooft expressed his view on the physics after the SM to include a new symmetry principle or new physical fields, (see in 't Hooft, 2005, section 12). The *property* derived in this book, the proof of *its conservation*, and the introduced **D** *field* are candidates.

The presented idea is based on the same facts like those considered in the SM, only it clusters the observations in another way (cf., another "version of facts"). Unlike existing alternative theories, for example, the supersymmetry (SUSY), which renders a new ("supersymmetric") brother to each particle, this model clusters the observed sources of fields in two-egg twin pairs, regarding them as isotopic states of each other. These field sources include the mass, the electric charge, and maybe others. A conserved property to appear as a consequence of their interaction, during which they can transform into each other's states, will predict new bosons that convey such a transformation. There is left "only" to observe the twin siblings of the SM bosons that mediate the interactions between the siblings of the SM field sources. In contrast to the SUSY, which renders fermion–boson pairs as new-born brothers to each other, the assumption argued in this book and earlier by Darvas (2009, 2011), renders fermion–fermion and boson–boson twins to each other. This radically new idea was called hypersymmetry (HySy).

This assumption does not assume new *fermions*; the twin brothers of fermions (and the mass) originate in splitting the existing ones. Fermions split as a result of

https://doi.org/10.1515/9783110713183-007

a newly interpreted property. The assumption is mathematically based (Darvas, 2009) and was proved in physical terms in Darvas (2011) and following lectures and papers. The second assumption was that the split field sources can interact with each other. The next condition is not only an assumption: it was introduced long ago but for long hypothetical velocity-dependent gauge field.

Bosonic twin brothers to be predicted should appear as the quanta of the **D** field (cf., Section 8.1) that mediate between the split fermion states, that means, between isotopic states of field-charges. The prediction of bosonic twin brothers will be discussed in Section 9.3.

The assumed theory on the interaction between field-charge siblings does not give an explanation for everything. It clusters known facts in another version, so it is a modest attempt to answer *a few* open questions of contemporary physics.

7.2 Introduction to the mathematics of the two simultaneous Noether currents in HySy

In order to retrospect, we remind that an old paper by Al-Kuwari and Taha (1991) discussed the conditions of local gauge invariance under a general non-Abelian group. They developed an interpretation of the Noether (1918) theorems applying the generalisation by Utiyama (1956, 1959), which imply the field equations for gauge vector fields and the existence of conserved Noether currents of global gauge invariance. They concluded that there were no extra conserved currents associated with local gauge invariance. We show that, in a more general case, there are further conserved Noether currents.

Noether's mathematical derivation allows the dependence of the concerned fields on any, general co-ordinate (cf., Darvas, 2018c). So do Al-Kuwari and Taha when they apply the same, general parameters as dependent variables of physical fields, however, they do not discuss the consequences of the application of the possible variables other than the four space–time co-ordinates. In this sense, they discuss the theorem in a restricted domain of variables. Such ignored but allowed and possible variables are, for example, the co-ordinates in the velocity four-space (configuration space). Indeed, certain fields may depend on the co-ordinates of the velocity space as well (cf., Section 6.2). In classical electrodynamics, for example, the force caused by the current of moving electric charges depends on the velocity of the charges with respect to the reference frame of the observer. Several descriptions of velocity-dependent phenomena handle the space–time co-ordinates as implicit variables.

We will reconstruct and extend an apparent clue, by Al-Kuwari and Taha, pertaining to the presence of a gauge field that depends on the co-ordinates of the velocity space (and implicitly on the space-time co-ordinates). The result will be an additional conserved Noether current that is localised in the velocity space. For the effects of a general non-Abelian group on the described local gauge invariance, we refer to R. Mills' (1989) review paper. We partially use the methods of his description of YM type gauge fields; however, we introduce a new type of localised gauge field that does not coincide with the YM field. We use the same letter (**D**) like Mills to denote a gauge field, but we denote another gauge field by **D** (cf., footnote 6.1). This **D** field has similar properties (the partial analogy explains the same notation), but it is *per definitionem* different from the field discussed by Mills (1989).

Concerning the interpretation of localisation, we must emphasise (before one would mistakenly claim that there is a possible loss of causality in our theory) that we do not use the term localisation restricted to the placement in space and time, rather in a generalised meaning, when we extend the role of the co-ordinates to a set of generalised variables (cf., Noether, 1918). These variables may be the four space–time co-ordinates or they may be others (and their number may vary). In her mathematical terms of invariant variational problems, the space–time co-ordinates did not play a distinguished role. According to her second theorem, other variables, such as velocity–space co-ordinates, are allowed too, which may implicitly depend on the space–time co-ordinates.

First, we will discuss the problem in the mathematical terms mentioned. Later we will discuss the possible physical consequences when we have derived the resulting equations. We draw the reader's attention again to the fact that in our discussion, the introduced **D** gauge field depends directly on the velocity–space co-ordinates, while the matter field depends directly on the four-dimensional (4D) space–time co-ordinates. In other words, this means that although we primarily used co-ordinates of the velocity–space, our derivations are indirect and include derivatives with respect to the space–time co-ordinates (cf., the introduction of the Λ_μ^ν tensor in the next subsection), and they will play important role in our conclusions. This is an expression of the fact that we observe the physical events (occurring even in the velocity–space) with respect to the 4D space–time.

One may now ask: Why do we take into consideration space–time co-ordinates in the form of implicit parameters (cf., also footnote 7.1)? There are pragmatic reasons. Generally, physics introduces dependence on the co-ordinates of a configuration space in the form$f(\dot{x}_\mu, x_\mu)$. In our case, this treatment would lead to eight parametric derivations, and large acceleration tensors, whose several elements would play no role in the discussion. For practical reasons, we could reduce the calculations to the most necessary ones by replacing the $f(\dot{x}_\mu, x_\mu)$ dependence with a $f(\dot{x}_\mu(x_\mu))$dependence. The localisation is present here too (in the mentioned generalised, Noetherian sense), although we are allowed another way of calculating it.

7.3 Noether's currents for gauge invariance localised in a velocity field

We discuss general, non-Abelian case. We are seeking for invariance between scalar fields and (gauge) vector fields that describe kinetic processes, the latter depending therefore primarily on velocity (Darvas, 2009). For this reason, we consider Lagrangians which depend on matter fields φ_k, and gauge fields $D_{\dot{\mu}\alpha}$ (dotted indices denote the velocity–space components), which all depend – in simple mathematical terms – on $x_{\mu,}$ given in this specific case by the formula: $D_{\dot{\mu}} = D_\mu(\partial x^\mu/\partial x_4)$ or in another form $D_{\dot{\mu}} = D_\mu[\dot{x}^\mu(x_\nu)]$, where $\dot{x}^\mu = \dot{x}^\mu(x_\nu)$; $(\mu, \nu = 1, 2, 3, 4)$.[7.1] We introduce a notation $\Lambda_\mu^\nu = \partial_\mu \dot{x}^\nu = \partial\dot{x}^\nu/\partial x_\mu$ for the Lorentz invariant acceleration that characterises the changes of the velocity–space components in the space–time. The *localisation* of **D** is taken into consideration in this way $D_\mu[\dot{x}^\mu(x_\nu)]$ (we refer to the generalised interpretation of localisation as defined in Section 7.2).[7.2]

This choice is in full accordance with the conditions set for Lagrangians in the second theorem of Noether.[7.3] The proof discusses general, non-Abelian case. The mathematical derivation is based on a transformation group G and the transformations of its elements into each other $T\left[G_{\infty,\rho}\right] = T\left[p_\alpha(x_\beta)\right]$, where the number of the parameters p and x are arbitrary, finite numbers $(\alpha = 1, \ldots, \rho)$; $(\beta = 1, \ldots, \sigma)$. The "$p$" are parameters on which the transformations, constituting the group elements,

---

7.1 We discussed in Section (6.2) that the presence of velocity dependence in several physical situations, when space–time dependence is considered only, as defined here, indirectly. The results to be derived (cf., Section 7.4) justify in an explicit form the assumption that a velocity-dependent field may have physical meaning. Let us see examples.

Consider the already discussed situation: We are taking measurements in a system of reference fixed to a lab. We are measuring effects of moving charges. The effects, for example, force, originating from the moving reference system of the charges, as causes of the measured effects, depend on first approximation on three sets of parameters – their space and time co-ordinates, their velocity components, and the charge. If we fix the values of one of those sets, the effects measured by us will no longer directly depend on those. For example, we can measure the effect originating from an electron. In this case, the value of the charge is given but it will no longer play the role of an independent variable. Let us imagine a more specific, but not rare, situation: we are measuring effects of a valence electron oscillating in the electromagnetic field between two atoms whose position is fixed in a crystal. In this case, although the electron executes motion, the change in its space co-ordinates is negligibly small compared to the distance from the measuring instrument in our lab. In practical terms its position does not change. The effect we are observing depends only on its actual velocity. Of course, we record its position in our lab, but we can consider this only as a dependent variable. We make use of this when we define the derivatives of the velocity in the reference system fixed to our lab.

7.2 Relativistic covariance under a Lorentz transformation $S(\Lambda)$ and its consequences are a standard part of quantum field theory textbooks for long, For example, Itsykson and Zuber (1980, section 2.1.3). Here we take into account the time derivatives of Lorentz transformed velocities.

7.3 In mathematical terms she did not specify either the physical-mathematical character or the number of applicable parameters (cf., Darvas, 2018c) .

depend. They take the form of functions $p_\alpha(x_\beta)$ and their derivatives. The group transformations depend on p and are finitely differentiable. G may take the form of various groups, depending on the concrete form of interaction in the subject, namely $SO(3,1)$, $U(1)$, $SU(2)$, $SU(3)$ in the case of the fundamental physical interactions.

Let the considered Lagrangian density be $L(\varphi_k, D_{\dot{\mu}\alpha})$, where φ_k, $(k=1, ..., n)$ denote so-called matter fields – which also include the velocity field $\dot{x}^\mu = \dot{x}^\mu(x_\nu)$ –, and $D_{\dot{\mu}\alpha}$, $(\alpha = 1, ..., N)$, denote (kinetic) gauge fields. We assume that $L(\varphi_k, D_{\dot{\mu}\alpha})$ is invariant under the local transformations of a compact, simple Lie group G generated by T_α, $(\alpha = 1, ..., N)$, where $\left[T_\alpha, T_\beta\right] = iC^\gamma_{\alpha\beta} T_\gamma$, and $C^\gamma_{\alpha\beta}$ are the so-called structure constants, corresponding to the actually considered individual physical interaction's symmetry group.[7.4,7.5] For simplicity, we assume that the matter fields belong to a single, n-dimensional representation of G.

Finally, we consider a local transformation $V(\dot{x}) \in G$ parameterised by $p_\alpha(\dot{x})$ that acts on ψ as $\psi = V\psi'$ in the form of

$$V(\dot{x}) = e^{-ip_\alpha(\dot{x})T_\alpha},$$

which is localised by $V = V[\dot{x}^\mu(x_\nu)]$. Please pay attention to the following important fact: we do not stipulate any restriction on the form of the Lagrangian density in question. $L(\varphi_k, D_{\dot{\mu}\alpha})$ may represent the Lagrangian density of any physical interaction that takes place in a matter field φ_k and a kinetic field $D_{\dot{\mu}\alpha}$. According to Noether's second theorem, these Lagrangians may depend on an arbitrary number of fields with an arbitrary number of derivatives by an arbitrary number of parameters.

The infinitesimal transformations of the matter and the gauge fields determine the covariant derivatives of ψ in the gauge field. (For invariance, we can require that the derivatives of ψ coincide with the derivatives of $V\psi'$). The infinitesimal transformations can be formulated as follows:

$$\delta\varphi_k = -ip_\alpha(\dot{x})(T_\alpha)_{kl}\varphi_l(\dot{x}) \ (k = 1, ..., n) \tag{7.1}$$

where the T_α are matrix-representation operators generating the group G, with the introduced commutation rule $\left[T_\alpha, T_\beta\right] = iC^\gamma_{\alpha\beta} T_\gamma$, and

$$\delta D_{\dot{\mu}\alpha} = \frac{1}{\lambda}\partial^{\dot{\rho}} p_\alpha(\dot{x})\partial_\mu \dot{x}^\rho + C^\gamma_{\alpha\beta} p_\beta(\dot{x}) D_{\dot{\mu}\gamma}(\dot{x}) \ (\alpha = 1, ..., N) \tag{7.2}$$

7.4 We partly follow the clues by Higgs (1966) and Weinberg (1972) at the beginning of their papers with the exception that we consider different dependencies in the potential and kinetic Hamiltonian terms.

7.5 For examples, in the case of $SU(2)$ symmetry, G consists of 2×2matrices with three independent components, representing a state doublet, and in the case of $SU(3)$ its matrix has eight independent components, representing a state triplet.

where $\partial^{\dot{\rho}} = \partial/\partial\dot{x}^{\rho}$, and ℷ (Hebrew g, gimel) denotes a general coupling constant, which can be replaced by a concrete coupling constant for each individual physical interaction.

For the induced infinitesimal transformation δL of the Lagrangian density $L(\varphi_k, D_{\dot{\mu}\alpha})$, on using the field equations for both the matter and the gauge fields, one obtains

$$\delta L = \partial_\mu \left(\frac{\partial L}{\partial(\partial_\mu \varphi_k)} \delta\varphi_k + \frac{\partial L}{\partial(\partial_\mu D_{\nu\alpha})} \delta D_{\nu\alpha} \right) \tag{7.3}$$

One would like to describe the events resulting in the interaction between the matter field and the kinetic (velocity-space dependent) gauge field, as they are observed from the usual 4D space–time. Therefore one needs to apply derivatives by the space–time co-ordinates. Substituting from (7.1) and (7.2) into (7.3), using the introduced notation $(\partial\dot{x}^\nu/\partial x_\mu) = \partial_\mu \dot{x}^\nu = \Lambda_\mu^\nu$ (Lorentz invariant acceleration), and executing a permutation of the indices, one can obtain

$$\delta L = \partial_\mu \left(\frac{\partial L}{\partial(\partial_\mu \varphi_k)} (-i) p_\alpha(\dot{x}) (T_\alpha)_{kl} \varphi_l(\dot{x}) \right) +$$
$$+ \partial_\mu \left(\frac{\partial L}{\partial(\partial_\mu D_{\dot{\nu}\alpha})} \cdot \frac{1}{ℷ} \partial^{\dot{\rho}} p_\alpha(\dot{x}) \Lambda_\nu^\rho \right) + \partial_\mu \left(\frac{\partial L}{\partial(\partial_\mu D_{\dot{\nu}\alpha})} C_{\alpha\beta}^\gamma \, p_\beta(\dot{x}) D_{\dot{\nu}\gamma}(\dot{x}) \right)$$

and from this

$$\delta L = \partial_\mu \left[\frac{\partial L}{\partial(\partial_\mu \varphi_k)} \frac{1}{i} (T_\beta)_{kl} \varphi_l(\dot{x}) - C_{\alpha\beta}^\gamma \frac{\partial L}{\partial(\partial_\mu D_{\dot{\nu}\alpha})} D_{\dot{\nu}\gamma}(\dot{x}) \right] p_\beta(\dot{x}) +$$
$$+ \begin{bmatrix} -\dfrac{\partial L}{\partial(\partial_\mu \varphi_k)} \dfrac{1}{i} (T_\beta)_{kl} \varphi_l(\dot{x}) \Lambda_\mu^\nu + C_{\alpha\beta}^\gamma \dfrac{\partial L}{\partial(\partial_\mu D_{\dot{\rho}\alpha})} D_{\dot{\rho}\gamma}(\dot{x}) \Lambda_\mu^\nu - \\ -\dfrac{1}{ℷ} \dfrac{\partial L}{\partial(\partial_\rho D_{\dot{\mu}\beta})} \partial_\rho \Lambda_\mu^\nu - \dfrac{1}{ℷ} \partial_\rho \left(\dfrac{\partial L}{\partial(\partial_\rho D_{\dot{\mu}\beta})} \right) \Lambda_\mu^\nu \end{bmatrix} \partial^{\dot{\nu}} p_\beta(\dot{x}) - \tag{7.4}$$
$$- \frac{1}{ℷ} \frac{\partial L}{\partial(\partial_\mu D_{\dot{\nu}\alpha})} \Lambda_\nu^\rho \partial_\mu \partial^{\dot{\rho}} p_\alpha(\dot{x})$$

It was observed (Al-Kuwari and Taha, 1991) that $p_\alpha(\dot{x})$, $\partial^{\dot{\mu}} p_\alpha(\dot{x})$, and $\partial_\mu \partial^{\dot{\nu}} p_\alpha(\dot{x})$, $(\alpha = 1, \ldots, N)$ are arbitrary and independent at any $\dot{x}$, so that the local gauge invariance of $L(\varphi_k, D_{\dot{\mu}\alpha})$, $\delta L = 0$ is equivalent to the following three conditions (7.5)–(7.7), according to the requirement to make the coefficients of $p_\alpha(\dot{x})$, $\partial^{\dot{\mu}} p_\alpha(\dot{x})$, and $\partial_\mu \partial^{\dot{\nu}} p_\alpha(\dot{x})$ equal to zero separately:

$$\partial_\mu \left[\frac{\partial L}{\partial(\partial_\mu \varphi_k)} \frac{1}{i} (T_\alpha)_{kl} \varphi_l(\dot{x}) - C_{\alpha\beta}^\gamma \frac{\partial L}{\partial(\partial_\mu D_{\dot{\nu}\beta})} D_{\dot{\nu}\gamma}(\dot{x}) \right] = 0 \tag{7.5}$$

$$\frac{\partial L}{\partial(\partial_\mu\varphi_k)}\frac{1}{i}(T_\alpha)_{kl}\varphi_l(\dot{x})\Lambda_\mu^\nu - C_{\alpha\beta}^\gamma\frac{\partial L}{\partial(\partial_\mu D_{\dot{\rho}\beta})}D_{\dot{\rho}\gamma}(\dot{x})\Lambda_\mu^\nu + \\ + \frac{1}{\lambda}\partial_\rho\left[\frac{\partial L}{\partial(\partial_\rho D_{\dot{\mu}\alpha})}\Lambda_\mu^\nu\right] = 0 \tag{7.6a}$$

or written in another form:

$$\frac{\partial L}{\partial(\partial_\mu\varphi_k)}\frac{1}{i}(T_\alpha)_{kl}\varphi_l(\dot{x})\Lambda_\mu^\nu - C_{\alpha\beta}^\gamma\frac{\partial L}{\partial(\partial_\mu D_{\dot{\rho}\beta})}D_{\dot{\rho}\gamma}(\dot{x})\Lambda_\mu^\nu + \\ + \frac{1}{\lambda}\frac{\partial L}{\partial(\partial_\rho D_{\dot{\mu}\alpha})}\partial_\rho\Lambda_\mu^\nu + \frac{1}{\lambda}\partial_\rho\left[\frac{\partial L}{\partial(\partial_\rho D_{\dot{\mu}\alpha})}\right]\Lambda_\mu^\nu = 0 \tag{7.6b}$$

$$\frac{1}{\lambda}\frac{\partial L}{\partial(\partial_\mu D_{\dot{\nu}\alpha})}\Lambda_\nu^\rho = 0 \tag{7.7a}$$

or considering that $\partial_\mu\partial^{\dot{\rho}}p_\alpha(\dot{x})$ is symmetric in μ and ρ:

$$\frac{1}{\lambda}\frac{\partial L}{\partial(\partial_\mu D_{\dot{\nu}\alpha})}\Lambda_\nu^\rho + \frac{1}{\lambda}\frac{\partial L}{\partial(\partial_\nu D_{\dot{\mu}\alpha})}\Lambda_\mu^\rho = 0 \tag{7.7}$$

Let us define

$$F_\alpha^{\mu\nu}(\dot{x}) = \frac{\partial L}{\partial(\partial_\mu D_{\dot{\rho}\alpha})}\Lambda_\rho^\nu \tag{7.8}$$

and

$$J_\alpha^{(1)\mu} = \lambda i\left[\frac{\partial L}{\partial(\partial_\mu\varphi_k)}(T_\alpha)_{kl}\varphi_l(\dot{x}) + C_{\alpha\beta}^\gamma\frac{\partial L}{\partial(\partial_\mu D_{\dot{\nu}\beta})}D_{\dot{\nu}\gamma}(\dot{x})\right] \tag{7.9}$$

$$j_\alpha^{(2)\nu} = \lambda i\left[\begin{array}{l}\frac{\partial L}{\partial(\partial_\mu\varphi_k)}(T_\alpha)_{kl}\varphi_l(\dot{x})\Lambda_\mu^\nu + C_{\alpha\beta}^\gamma\frac{\partial L}{\partial(\partial_\mu D_{\dot{\rho}\beta})}D_{\dot{\rho}\gamma}(\dot{x})\Lambda_\mu^\nu - \\ -\frac{1}{\lambda}\frac{\partial L}{\partial(\partial_\rho D_{\dot{\mu}\alpha})}\partial_\rho\Lambda_\mu^\nu\end{array}\right] \tag{7.10}$$

Then, based on (7.5')

$$\partial_\nu J_\alpha^{(1)\nu}(\dot{x}) = 0 \tag{7.9a}$$

(a) Using (7.5) and (7.6b) as well as (7.7a) and taking into consideration the condition that the tensor Λ_μ^ν is not identically zero in all its elements, one obtains the following:

$$J_\alpha^{(1)\nu}(\dot{x}) = \partial_\mu\frac{\partial L}{\partial(\partial_\mu D_{\dot{\nu}\alpha})}$$

that is

$$j_\alpha^{(2)\nu}(\dot{x}) = J_\alpha^{(1)\mu}(\dot{x})\Lambda_\mu^\nu \tag{7.11}$$

One can observe that (7.11) mixes the components of the gauge-field currents $J_\alpha^{(1)\mu}$ and $j_\alpha^{(2)\nu}$ depending on the 4D velocity space in a similar way like the Lorentz transformation mixes the co-ordinates of four-vectors in the 4D space–time. Note that the Λ_μ^ν tensor characterises the changes of the velocity–space components in the space–time.

(b) (7.9a) and (7.11) together with (7.7) yield that $J_\alpha^{(1)\nu}(\dot{x})$ is a conserved Noether current.

In order to admit this, let us consider $F_\alpha^{(1)\mu\nu}$ in the form

$$F_\alpha^{(1)\nu\mu}(\dot{x}) = \frac{\partial D_{\nu\alpha}(\dot{x})}{\partial \dot{x}_\mu} - \frac{\partial D_{\mu\alpha}(\dot{x})}{\partial \dot{x}_\nu} - i\lambda C^{\ \gamma}_{\alpha\beta} D_{\dot{\nu}\beta}(\dot{x}) D_{\dot{\mu}\gamma}(\dot{x})$$

Considering again (7.7) and (7.7a) $J_\alpha^{(1)\nu}(\dot{x})$ satisfies

$$J_\alpha^{(1)\nu}(\dot{x}) = \partial_\mu F_\alpha^{(1)\mu\nu}(\dot{x}) = \hat{\partial}_\mu F_\alpha^{(1)\mu\nu}(\dot{x}) - i\lambda C^\gamma_{\alpha\beta} D_{\dot{\mu}\beta}(\dot{x}) \times F_\gamma^{(1)\mu\nu}(\dot{x}) \tag{7.9b}$$

which demonstrates the covariant form of $J_\alpha^{(1)\nu}(\dot{x})$.[7.6]

Substituting $J_\alpha^{(1)\mu}$ in (7.5), as well as in (7.6a) and taking into consideration (7.7a) with the condition that all elements of the tensor Λ_μ^ν are not zero, one obtains

$$\partial_\nu J_\alpha^{(1)\nu} = 0 \tag{7.12}$$

and

$$J_\alpha^{(1)\mu} = \partial_\rho \left(\frac{\partial L}{\partial(\partial_\rho D_{\dot{\mu}\alpha})} \right)$$

We see also that

$$J_\alpha^{(1)\nu} = \partial_\mu F_\alpha^{(1)\mu\nu} \tag{7.13}$$

(7.12) and (7.13) – together with (7.7) – mean that $J_\alpha^{(1)\nu}$ is a conserved Noether current.

Now, let us investigate what conditions do the two currents, namely $J_\alpha^{(1)\mu}$ and $j_\alpha^{(2)\nu}$, fulfil, and if there are any, which invariance conditions are those?

7.6 Careted $\hat{\partial}$ denotes covariant derivative, here and in the following.

(c) Based on (7.7), one can define also

$$F_\alpha^{\mu\nu}(x) = \frac{\partial L}{\partial(\partial_\mu D_{\dot{\rho}\alpha})}\Lambda_\rho^\nu$$

and

$$j_\alpha^{(2)\nu}(x) = \lambda i \left[\begin{array}{l} \frac{\partial L}{\partial(\partial_\mu \varphi_k)}(T_\alpha)_{kl}\varphi_l(\dot{x})\Lambda_\mu^\nu + C_{\alpha\beta}^\gamma \frac{\partial L}{\partial(\partial_\mu D_{\dot{\rho}\beta})} D_{\dot{\rho}\gamma}(\dot{x})\Lambda_\mu^\nu - \\ -\frac{1}{\lambda}\frac{\partial L}{\partial(\partial_\rho D_{\dot{\mu}\alpha})}\partial_\rho \Lambda_\mu^\nu \end{array} \right]$$

Note that due to the indirect derivation $F_\alpha^{\mu\nu}$ and $j_\alpha^{(2)\nu}$ – in contrast to $F_\alpha^{(1)\mu\nu}$ and $J_\alpha^{(1)\nu}$ – are functions of x instead of $\dot{x}$.

Substituting $j_\alpha^{(2)\nu}$ in (7.6b) and taking again into consideration (7.7a) with the condition that all elements of the tensor Λ_μ^ν are not zero, one can obtain that

$$j_\alpha^{(2)\nu} = \partial_\mu F_\alpha^{\mu\nu} \tag{7.14}$$

The divergence of $j_\alpha^{(2)\nu}$: $\partial_\nu j_\alpha^{(2)\nu} = \partial_\nu \partial_\mu F_\alpha^{\mu\nu}$ does not vanish identically. In accordance with the requirement that the derivatives of all ψ on which the operators are effective shall coincide with the derivatives of their transformed (by V) states; let us define $\alpha = 1, \dots, N$ contravariant forms $F_\alpha^{(2)\mu\nu}$–s so that their covariant derivatives be equal to the derivatives of the $F_\alpha^{\mu\nu}$–s:

$$\hat{\partial}_\mu F_\alpha^{(2)\mu\nu}(x) = \partial_\mu F_\alpha^{\mu\nu}(x) \tag{7.15}$$

where, as defined earlier, the careted $\hat{\partial}_\mu$ denotes the covariant derivative and the covariant derivative of $F_\alpha^{(2)\mu\nu}$ can be written as[7.7]

$$\hat{\partial}_\mu F_\alpha^{(2)\mu\nu}(x) = \partial_\mu F_\alpha^{(2)\mu\nu} + i\lambda C_{\alpha\beta}^\gamma D_{\dot{\omega}\beta}\Lambda_\mu^\omega \times F_\gamma^{(2)\mu\nu} \tag{7.16}$$

7.7 Note the following: The YM theory (Yang and Mills, 1954; Mills, 1989) introduced the covariant form of $F_\alpha^{(2)\mu\nu}$ derived from the Lagrangian density of a specific fermion field. We do not make any preliminary assumption concerning the Lagrangian density of the field. We defined the covariant $F_\alpha^{(2)\mu\nu}$ in an essentially different, independent way, based on the requirement of their invariant transformation, and thus it got rid of any specific form of the Lagrangian density. The importance of this different approach becomes apparent looking at the discussion of the results by R. Mills himself (1989) in the light of the theory of fibre bundles. He observed that the applied covariant derivatives bear a very close relationship to the covariant derivatives of general relativity theory, and the quantities $F_\alpha^{(2)\mu\nu}$ demonstrate close analogy to the curvature tensor of general relativity. Since YM theory derived $F_\alpha^{(2)\mu\nu}$ from a specific form for the Lagrangian density, they could not state anything more than an observed similarity. Furthermore, the Lagrangian-invariant introduction of $F_\alpha^{(2)\mu\nu}$ and their covariant derivatives also leaves free the opportunity for application to gravitational fields. An advantage of this treatment is to find conserved Noether currents that are of an identical-form in various gauge fields.

The following $F_\alpha^{(2)\mu\nu}$ fulfil the requirement formulated in (7.16):

$$F_\alpha^{(2)\mu\nu}(x) = \frac{\partial D_{\dot\rho\alpha}\Lambda_\mu^\rho}{\partial x_\nu} - \frac{\partial D_{\dot\sigma\alpha}\Lambda_\nu^\sigma}{\partial x_\mu} - i\lambda C_{\alpha\beta}^\gamma D_{\dot\rho\beta}\Lambda_\mu^\rho D_{\dot\sigma\gamma}\Lambda_\nu^\sigma \tag{7.17}$$

This covariant $F_\alpha^{(2)\mu\nu}$ transforms under a non-Abelian, velocity-dependent gauge transformation in the same way as the isovector $F_\alpha^{\mu\nu}$ does.[7.8] Thus one can replace the divergence of $F_\alpha^{\mu\nu}$ by the divergence of $F_\alpha^{(2)\mu\nu}$.

Applying (7.16) then (7.14)

$$\partial_\mu F_\alpha^{(2)\mu\nu}(x) = \partial_\mu F_\alpha^{\mu\nu}(x) - i\lambda C_{\alpha\beta}^\gamma D_{\dot\omega\beta}(\dot x)\Lambda_\mu^\omega \times F_\gamma^{(2)\mu\nu}(x)$$

or (7.18)

$$\partial_\mu F_\alpha^{(2)\mu\nu}(x) = j_\alpha^{(2)\nu}(x) - i\lambda C_{\alpha\beta}^\gamma D_{\dot\omega\beta}(\dot x)\Lambda_\mu^\omega \times F_\gamma^{(2)\mu\nu}(x)$$

(7.18) defines a current

$$J_\alpha^{(2)\nu}(x) = \partial_\mu F_\alpha^{\mu\nu}(x) - i\lambda C_{\alpha\beta}^\gamma D_{\dot\omega\beta}(\dot x)\Lambda_\mu^\omega \times F_\gamma^{(2)\mu\nu}(x) \tag{7.19}$$

or, which is the same:

$$J_\alpha^{(2)\nu}(x) = \partial_\mu F_\alpha^{(2)\mu\nu}(x) \tag{7.20}$$

whose 4D divergence is automatically zero, due to (7.15) and (7.16).

7.4 Discussion of the mathematical results

Referring to (7.12), (7.13) and (7.15), (7.16), (7.20) we have obtained the following two sets of equations:

$$J_\alpha^{(1)\nu} = \partial_\mu F_\alpha^{(1)\mu\nu} \qquad \partial_\nu J_\alpha^{(1)\nu} = 0 \tag{7.21}$$

$$J_\alpha^{(2)\nu} = \partial_\mu F_\alpha^{(2)\mu\nu} \qquad \partial_\nu J_\alpha^{(2)\nu} = 0 \tag{7.22}$$

Completed with (7.7) this set demonstrates, that in the presence of a velocity-dependent gauge field, there exist two (families of) conserved Noether currents.

Although the two conserved currents are not independent (cf., (7.11)) in the presence of a kinetic gauge field they exist simultaneously. In the absence of the velocity-dependent gauge field, we get back to the results based on calculations in

7.8 Weinberg (1996) noted that $\hat\partial_\mu\psi$ transforms just like ψ itself.

a space–time dependent gauge field. Thus without employing accelerations, we derived the same conserved currents. At rest or at relatively low velocities, the second Noether current vanishes and we have only the classically known conserved current. However, at high velocities, in the presence of a kinetic field, there appears a parallel conserved current: we step beyond the limits of the SM (cf., Section 10.3.3). This result justifies our preliminary assumption that handling the space–time coordinates as implicit parameters not only provides additional information but also preserves the physical relevance of the theory.

Taking into account the conditions how we have obtained these currents, one can write $J_\alpha^{(1)\mu}$ as

$$J_\alpha^{(1)\nu}(\dot{x}) = \lambda i \frac{\partial L}{\partial(\partial_\nu \varphi_k)} (T_\alpha)_{kl} \varphi_l(\dot{x}) \tag{7.23}$$

This form coincides with the usual conserved Noether currents known in field theories.

The most significant conclusion of the presented derivation (cf., Darvas, 2009) is that in the presence of a velocity-dependent gauge field **D**, there appear extra $J_\alpha^{(2)\nu}$ conserved currents. Taking into account (7.6b), (7.10), and (7.11), then (7.7a), one can now write (7.19) in the form

$$J_\alpha^{(2)\nu}(x) = \lambda i \left[\frac{\partial L}{\partial(\partial_\mu \varphi_k)} (T_\alpha)_{kl} \varphi_l(\dot{x}) \Lambda_\mu^\nu - C_{\alpha\beta}^\gamma D_{\dot{\omega}\beta}(\dot{x}) \Lambda_\mu^\omega \times F_\gamma^{(2)\mu\nu}(x) \right] \tag{7.24}$$

where $F_\alpha^{(2)\mu\nu}(x)$is

$$F_\alpha^{(2)\mu\nu}(x) = \frac{\partial D_{\dot{\rho}\alpha} \Lambda_\mu^\rho}{\partial x_\nu} - \frac{\partial D_{\dot{\sigma}\alpha} \Lambda_\nu^\sigma}{\partial x_\mu} - i\lambda C_{\alpha\beta}^\gamma D_{\dot{\rho}\beta} \Lambda_\mu^\rho D_{\dot{\sigma}\gamma} \Lambda_\nu^\sigma$$

Their dependence on the velocity–space gauge is apparent, although, none of the conserved vector currents involves the gauge parameters $p_\alpha(\dot{x})$ and its derivatives. From (7.24) with (7.23) [and (7.11), (7.15)] we see that $J_\alpha^{(1)\nu}$ and $J_\alpha^{(2)\nu}$ exist simultaneously.

From (7.21) and (7.23), considering (7.13), one obtains

$$\partial_\mu F_\alpha^{(1)\mu\nu}(\dot{x}) = \lambda i \frac{\partial L}{\partial(\partial_\nu \varphi_k)} (T_\alpha)_{kl} \varphi_l(\dot{x}) \tag{7.25}$$

From (7.22) and (7.24), considering (7.18) and (7.16), one obtains

$$\hat{\partial}_\mu F_\alpha^{(2)\mu\nu}(x) = \lambda i \frac{\partial L}{\partial(\partial_\mu \varphi_k)} (T_\alpha)_{kl} \varphi_l(\dot{x}) \Lambda_\mu^\nu \tag{7.26}$$

7.5 Physical considerations

Relations (7.25) and (7.26) provide the equations of motion for the potential part[7.9] of the system's Lagrangian density. According to a note by Brading and Brown (2000), based on an observation of Utiyama (1959, p. 27), generally, the case is that when (7.25) or (7.26) is satisfied, the matter-field current associated with the Lagrangian acts as the source for the gauge fields. The note (Brading and Brown, 2000; Utiyama, 1959) is a consequence of the fact that the matter-field dependent and the gauge-field dependent currents are at separate sides in each of the latter two equations.[7.10] Here the only condition assumed was that the field equations be satisfied. No restriction was imposed on the form of the Lagrangian density except that it be invariant under local gauge transformations as defined in (7.3). The covariant dependence on the velocity-space gauge field is obvious from (7.26), and we showed it in a similar way for (7.25) in (7.9b). *The derived conserved currents make a correspondence between the* matter fields *and the* kinetic *(velocity-dependent)* gauge fields. *They open the way to conclude an invariance between the sources of the scalar fields on the one side and the gauge vector fields on the other.*

One may be concerned that the applied conditions could contradict the usual physical theories. We show that such anxiety does not hold. In a boundary situation, namely in the absence of a velocity-dependent gauge field – an identical situation to the one originally studied by Al-Kuwari and Taha (1991) – we obtain the same currents that were derived by them in a space–time dependent field, cf., (7.21) and (7.23). In other words, in the absence of relativistically high velocities or acceleration, the effect of the velocity-dependent gauge field can be neglected, and we get back to the same currents as derived in the former, cited work [cf., (7.23), and compare it with the (7.24) derived here]. This boundary result justifies our choice and method. Nevertheless, in the presence of a velocity-dependent gauge field, we derived new conserved Noether currents (7.22) and (7.24).

In short, the *novelty in our treatment* is twofold: One is a merely mathematical-technical one, that is, the implicit parametrisation of the space–time co-ordinates. The other is the consideration of a gauge field **D** that depends on the temporal derivatives of the space–time co-ordinates; and that these may have physical meaning. The *novelty in our results* is the derivation of *new* conserved currents.

7.9 That is, which serves as the source for the characteristic field-charges of the given fields.

7.10 Brading and Brown (2000) call the equations defined in (7.21)–(7.26) "coupled field equations", after the form of the connection that they describe between the different (matter and gauge) fields appearing in the Lagrangian. This holds always when the Euler–Lagrange equations are assumed to be satisfied for all the fields on which the Lagrangian depends (or more precisely, for all the fields whose transformations depend on $\partial^{\dot{\mu}} p_\alpha$).

8 CONSERVATION OF THE ISOTOPIC FIELD-CHARGE SPIN (IFCS)

It is easy to see that $F_{\alpha}^{(1)\mu\nu}(\dot{x})$ and $F_{\alpha}^{(2)\mu\nu}(x)$ transform in the same way, as isovectors, under the local transformation $V(\dot{x}) \in G$:

$$F_{\alpha}^{(1)'\mu\nu}(\dot{x}) = V^{-1}F_{\alpha}^{(1)\mu\nu}(\dot{x})V \quad \text{and} \quad F_{\alpha}^{(2)'\mu\nu}(x) = V^{-1}F_{\alpha}^{(2)\mu\nu}(x)V.$$

Notice that the forms of $J_{\alpha}^{(1)\nu}(\dot{x})$ conserved currents, in the presence of velocity depending fields, coincide with the form of those currents that we had obtained for space–time depending fields. With respect to this identical form, as well as to the variety of the symmetry groups that they may obey, one can replace $\varphi(\dot{x}) \to \varphi(x)$, $D(\dot{x}) \to B(x)$, and $J_{\alpha}^{(1)}(\dot{x}) \to j_{\alpha}^{(1)}(x)$, where $B(x)$ denotes familiar physical gauge fields with symmetries, for example, $U(1)$, $SU(2)$, [and $SU(2) \times U(1)$], $SU(3)$ or $SO(3,1)$, with the substitution of ℷ by the corresponding coupling constants. $F_{\alpha}^{(1)\mu\nu}(\dot{x})$ take the same forms and transform in a velocity-dependent **D** gauge field like the components of a $j^{\nu}(x)$current and isovectors $f^{\mu\nu}(x)$ of a general matter field $\varphi(x)$ and gauge field **B**, defined by $f^{\mu\nu} = \partial^{\nu}B_{\mu} - \partial^{\mu}B_{\nu} - ℷ B_{\mu} \times B_{\nu}$ in the four-dimensional space–time. (This yields the information that in a boundary situation, i.e. in the absence of relativistic accelerations, our derivation produces the same result as it was known without the assumption of a velocity-dependent gauge field. We got back to the results that were known in the absence of a velocity-dependent gauge field, and that were based on calculations in an only space–time dependent gauge field. So, without employing accelerations, we derived the same conserved currents. This justifies our preliminary assumption that handling the space–time coordinates as implicit parameters not only provides additional information but also preserves the physical relevance of the theory.)

8.1 First conserved quantity: conservation of the field-charge (ℸ)

In a general case, the T_{γ} (that appear in the currents), as introduced earlier, are matrix-representation operators generating the group G, with the mentioned commutation rule $[T_{\alpha}, T_{\beta}] = iC_{\alpha\beta}^{\gamma'}T_{\gamma}$. They can be replaced by concrete operators of the concerned fields, according to their characteristic symmetry groups, like $U(1)$, $SU(2)$, $SU(3)$, or $SO(3,1)$, and ℷ can be substituted by the concrete coupling constants of the individual physical fields. Thus, in a general case, and with group G of an arbitrarily chosen physical field in SM, denoted by **B**, one can write $\varphi(x)$ and **B** in the equations for the currents $J_{\alpha}^{(1)\mu}$ and substitute the earlier equations derived in Section 7 with

https://doi.org/10.1515/9783110713183-008

$$J_\alpha^{(1)\nu}(x) = \lambda i \frac{\partial L}{\partial(\partial_\nu \varphi_k)}(T_\alpha)_{kl}\,\varphi_l(x), \quad J_\alpha^{(1)\nu}(x) = \partial_\mu F_\alpha^{(1)\mu\nu}(x)$$

$$F_\alpha^{(1)\mu\nu}(x) = \frac{\partial L}{\partial(\partial_\mu B_{\nu,\alpha}(x))}$$

and (8.1)

$$\hat{\partial}_\mu F_\alpha^{(1)\mu\nu}(x) = \partial_\mu F_\alpha^{(1)\mu\nu}(x) + i\lambda C^\gamma_{\alpha\beta} B_{\mu,\beta}(x) \times F_\gamma^{(1)\mu\nu}(x)$$

The operators of the quanta of the given physical field are determined by the generators $\{T_\alpha\}$ of the symmetry group of the respective field. The full conserved field-charge currents $J_\alpha^{(1)\mu}$ will provide the conserved quantities of the field $\varphi(x)$, which the gauge field **B** interacts with. We called these conserved quantities field-charges and denoted them by ך. We can get the conserved quantity by the integration of the current in the usual way, applying Gauss' theorem, where the integral of the spatial components vanishes at an infinite boundary, and we get the following:

$$\frac{\mathrm{d}}{\mathrm{d}t}\frac{\lambda}{c}\int \frac{\partial L}{\partial(\partial_4 \varphi_k)}(T_\alpha)_{kl}\,\varphi_l(x)\mathrm{d}V = 0 \tag{8.2}$$

where the integral provides the conserved field-charge ך of the source field φ.

The results derived in this subsection coincide with the well-known conservation laws of field theories. We treat it here in order to make it comparable with the results of the Section 8.2 and to demonstrate that the two conserved quantities appear simultaneously (Section 8.3).

8.2 Second conserved quantity: conservation of the isotopic field-charge spin (Δ)

Similarly $J_\alpha^{(1)\mu} - J_\alpha^{(2)\nu}(x)$ are also conserved and yield a conservation law. We will denote the conserved quantity derived from $J_\alpha^{(2)\nu}(x)$ by Δ.

The conserved current in the kinetic field can be read from (7.24). The right-hand side of (7.24) represents the full conserved current of Δ, which includes the contribution of the **D** field.[8.1]

We have introduced the **D** field – a kinetic field to be responsible for the expected transformation of the isotopic field-charges (IFC) into each other – to counteract the

8.1 Similar attempts (like ours in the velocity space) were made by Pons, Salisbury and Shepley (2000) in the phase space (with a particular mapping from the configuration space to phase space), and they anticipated the quantization of the models.

dependence of the V transformation on $\dot{x}_\mu$. The field equations, which are satisfied by the 12 independent components of the **D** field and their interaction with any field that carries conserved quantities, are unambiguously determined by the defined currents and covariant $F^{(2)\mu\nu}$-s constructed from the components of **D**. Considering a general Lorentz- and gauge-invariant Lagrangian, we obtain from the equations of motion that $J^{(2)1,2,3}$ and $J^{(2)4}$ are, respectively, the *current density* and the *density* of Δ. The total amount of the conserved quantity represented by the $J_\alpha^{(2)\nu}(x)$current in the **D** field,

$$\Delta = \frac{i}{\lambda}\int J^{(2)4}d^3x \tag{8.3}$$

is independent of time and invariant under Lorentz transformation. $J^{(2)\nu}$ does not transform as a vector, while Δ transforms like a vector under rotations in its field.

Associated with the derived two conserved currents, we found two conserved quantities ℸ and Δ in the fields φ and **D**, respectively. ℸ was identified with the field-charges. We are due to interpreting the physical meaning of Δ (Section 8.4).

8.3 Coupling of the two conserved quantities (ℸ and *Δ*)

We were made acquainted with the notion of the IFC ($ℸ_V$ and $ℸ_T$) in Section 4. There, we expected them to be able to transform into each other, but we had no evidence for the mathematical description of this conjecture. In Section 5, we demonstrated that the group of hypersymmetry (HySy) isomorphic with $SU(2)$, represented by the τ algebra, should describe such a transformation. In order to realise this option, one needs a conserved quantity conveyed by a conserved current. Such conserved quantity and current must be interpreted in a specific field. Provided that the IFC can transform into each other, they *should possess a property* which is conserved during that transformation. We show the existence of this property and identify it with Δ in the next (8.4) subsection. In this order, we demonstrate that the derived two conserved currents and field-charges are coupled, moreover, the introduced **D** field is just that in which the sought property of the IFC exists and that justifies to identify it with the conserved quantity derived from the current $J_\alpha^{(2)\nu}$. In this case, the quanta of the **D** field may carry the sought conserved property of the IFC.

The dependence of the two currents $J_\alpha^{(1)\mu}$ and $J_\alpha^{(2)\mu}$ on each other [cf., eqs. (7.11), (7.20), and (7.14)] has physical consequences. First, it justifies that the quantities, whose conservation they represent and which are coupled (by $\Lambda_\mu^\nu = \partial_\mu \dot{x}^\nu$), exist simultaneously. Second, the coupling of a conserved quantity in a space–time dependent field – which coincides with one of our familiar physical SM fields – with another in a kinetic (velocity-dependent) gauge field indicates that *the derived conservation verifies* just *the invariance between two isotopic states of the field-charges, namely between the potential* $ℸ_V$ *and the kinetic* $ℸ_T$, what we intended to prove. (Remember that ℸ can be field-charges of various physical fields marked in common with **B**, while Δ represents a single quantity belonging to the kinetic gauge field **D**.)

8.4 Interpretation of the isotopic field-charge spin (Δ)

The physical meaning of Δ can be understood, when we specify the transformation group associated with the **D** field, which describes the transformations of $\urcorner$ (i.e. the field-charges). We saw that $\urcorner$ can take two (potential and kinetic) isotopic states $\urcorner_V$ and $\urcorner_T$ in a simple unitary abstract space. Their symmetry group – what we called HySy – is *SU*(2) isomorphic (cf., Section 5.2.5, Theorem 8) that can be represented by 2×2 T_α matrices. There are three independent T_α that may transform into each other, following the rule $[T_\alpha, T_\beta] = iC^\gamma_{\alpha\beta} T_\gamma$, where the structure constants can take the values $0, \pm 1$. Let T_1 and T_2 be those which do not commute with T_3; they generate transformations that mix the different values of T_3, while this "third" component's eigenvalues represent the members of a Δ doublet. For the IFC compose a $\urcorner$ doublet of $\urcorner_V$ and $\urcorner_T$, and the field's wave function can be written as

$$\psi = \begin{pmatrix} \psi_T \\ \psi_V \end{pmatrix} \tag{8.4}$$

Equation (8.4) is the wave function for a single particle which may be in the "potential state", with amplitude ψ_V, or in the "kinetic state", with amplitude ψ_T. ψ in eq. (8.4) represents a mixture of the potential and kinetic states of the $\urcorner$, and there are T_α that govern the mixing of the components ψ_V and ψ_T in the transformation. T_α ($\alpha = 1, 2, 3$) are representations of operators which can be identified with the three components of Δ: $\Delta_1, \Delta_2, \Delta_3$ that follow the same (non-Abelian) commutation rules as do the T_α matrices, $[\Delta_1, \Delta_2] = i\Delta_3$, and so on. These operators represent the charges of a gauge field, and ψ are the fields on which the operators of the gauge fields act.

We obtained that *in the presence of kinetic fields we got two conserved currents that are effective simultaneously*. The kinetic gauge field **D** is present simultaneously with the interacting matter [φ] and gauge [**B**] fields. The presence of **D** corresponds to the property of the field-charges $\urcorner$ of the individual fields that they split into two isotopic states. As we saw, the identification of the components of Δ with generators of the HySy group is a consequence of the coincidence in their transformation rules. In the sense of its transformation rules, Δ represents a spin-like property.

In contrast to the spin that transforms into the space and to the isotopic spin that transforms in the YM **B** field, Δ transforms in the **D** field. According to the similar[8.2]

8.2 What is the essence of this similitude? It is in the symmetry group of their transformations. The *SU*(2) group of the isotopic spin transforms 2+2 spinor quantities into each other. The group of the transformation properties of the spin, the isotopic spin and weak isospin – analogously – we give the name to the two states that the property $\varDelta$ can occupy "*isotopic field-charge spin*" (IFCS). The IFCS is the sought property of the IFC. Due to the Noether current $J_{\alpha}^{(2)\mu}$, it is conserved. The source of the IFCS ($\varDelta$) is the field $\varphi(\dot{x})$, in interaction with the kinetic gauge field **D**.

The quanta of the **D** field should carry IFCS $\varDelta$. The $\varDelta$ doublet, as a conserved quantity, is related to the two isotopic states of field-charges (ℸ), and the associated operators ($\varDelta_{\mathrm{i}}$) induce transitions from one member of the doublet to the other. We discuss the quanta of the **D** field in Section 9.3.

The latter three paragraphs defined the IFCS (Δ). We proved its conservation afore. The conservation of Δ has been associated with the group of HySy.

HySy – which is also $SU(2)$ compatible – transforms coefficients of three-vector components into one-component scalars and back.

9 ISOTOPIC FIELD-CHARGES IN FUNDAMENTAL INTERACTIONS

We proved in Section 7.5 that isotopic field-charges (IFC) own a conserved property. We called this property, based on certain formal similitude, isotopic field-charge spin (IFCS), and denoted it by Δ. Δ exists parallel with the earlier known properties of the field-charges of the respective fields, provided that there is a velocity-dependent (kinetic) gauge field present. While deriving the conservation of Δ, we used a Lagrangian density that we have not specified. In fact, the derivation worked and its results hold without prescribing any restriction for the form of the Lagrangian.

The notion of IFC ($\daleth_V$ and $\daleth_T$) was defined for the electromagnetic and the gravitational fields in Section 4.2. The derivation of a Lagrangian-independent, universal conserved current ($J^{(2)\nu}$) and respective conserved property (Δ) allows to extend the notion of the IFCS to any field that can be characterised by a Lagrangian that appears in $J_{\alpha}^{(2)\nu}$, satisfies equations of motion and the field equations, as well as is invariant under the infinitesimal transformation obtained in Eq. (7.3) in Section 7.3. All field-charges of fields that can be characterised by such Lagrangians are open to being involved under the force of the IFC theory.

These considerations allow us to assume that the strong and the weak field-charges split in isotopic twins as well. How can one interpret them?

9.1 Isotopic field-charges in strong and electroweak interactions

Certain similitude between the relation and different physical roles of gravitational mass and inertial mass on the one hand, and of Coulomb charges and kinetic charges on the other, seems obvious. They are field-charges of different kind fields; however, the gravitational mass and the electric Coulomb charge serve both as sources of their respective fields ($\daleth_V$), associated with a central, scalar potential, and represent bound states of the respective field-charges. At the same time, the inertial mass and the kinetic electric charge are both associated with the kinetic part of the Hamiltonian of their interaction field ($\daleth_T$), both appear in vector components of currents and are the sources of momentum-like forces, namely of the momentum and the magnetic momentum, and both represent free states of the respective field-charges.

We cannot observe the same doublets in the case of colours, the charges and sources of the *strong field*. Since quarks, the carriers of the colours are confined in the baryons, it is hard to detect their free states (only among products of high energy collisions), and we can conclude the existence of such states only with the assumption of the asymptotic freedom. We may assume that if one detects unconfined quarks, one can observe similar bound and free, potential and kinetic, scalar and

https://doi.org/10.1515/9783110713183-009

vector property couples in their colour charge behaviour. Hard radiation of quarks, which is a rare phenomenon compared to the soft radiation, can change the overall flow of the energy and momentum to a large extent (Wilczek, 2003). This situation corresponds to an (asymptotically) free state of fast-moving quarks, which represent current-like kinetic states of the carrier of the strong field-charge. However, confined quarks cannot move very fast for long due to the limited diameter of a baryon. So their interactions involve soft radiation, which is detected in the majority of cases. The more common soft radiation allows concluding the bound states of the carriers of the colours that should be scalars and represent the sources of the strong field. Considering hard and soft radiation, we assume that (all the three) colour charges appear in two isotopic states as well.

Electroweak interaction is subject of combined $[U(1) \times SU(2)]$ symmetries. Electroweak interaction effects the particle families (the various flavours) of quarks and leptons, as the interaction field's sources (field-charges). Their, not always perfect[9.1] symmetries ensure the conservation laws of their field-charges. The material carriers (potential sources) of these electroweak interactions, over their mass and electric charge, have conserved field-charges (as quantum numbers). Such conserved properties are *weak isospin* and *flavour* in weak neutral current effects (in generalized Weinberg–Salam models), and there are conserved also the baryon number and lepton number of the carriers of the flavours, weak hypercharge (as a combined property), which are all conserved under special conditions[9.2] (cf., Table 9.1 in Section 9.2). The weak charges as sources of the weak field appear both in their scalar (potential) and in their vector (kinetic) form in the so-called $\boldsymbol{B}$ (for $U(1)$ symmetric) and $\boldsymbol{W}$ (for $SU(2)_L$ symmetric) fields of the electroweak interaction's Lagrangian. So, we have well-founded reason to assume that the weak charges (quarks and leptons in various flavours) are split into IFC twins too.

At least, our experience obtained from masses and electric charges, weak charges, and the indirect observations on strong colours allow us to assume a field-independent property – what we called IFC – which is responsible for similar, double behaviour. We assume IFC to be a property that can be attributed to field-charges of any known interaction field.

9.1 According to Weinberg (1997) "a spontaneously broken exact symmetry does not look at all like an approximate symmetry".

9.2 Glashow and Weinberg (1977) explored the consequences of the assumption that the direct and induced weak neutral currents in an $SU(2) \otimes U(1)$ gauge theory conserve all quark flavours naturally, that is, for all values of the parameters of the theory. This requires that all quarks of a given charge and helicity must have the same values of weak T_3 and $\vec{T}^2$ (T denotes here the weak isospin). If all quarks have charge + 2/3 or −1/3, the only acceptable theories are the "standard" and "pure vector" models, or their generalizations to six or more quarks. In addition, there were severe constraints on the couplings of Higgs bosons, which apparently could not be satisfied in pure vector models.

9.2 Summary: field-charges in all the four fundamental interactions

The substance of the obtained results is justified by the ability of the two kinds of charges of the electromagnetic field, the two kinds of the charge of the gravitational field, field-charges of the electroweak interaction, and presumably of the strong field, respectively, to be transformed into each other by a gauge transformation of hypersymmetry (HySy) in the velocity-dependent **D** field.

This gauge transformation involved the existence of a conserved property that we can summarise in the following way:

- Since the transformation affects the *isotopic states* of the individual *field-charges* (ℸ) in all standard model (SM) interactions, this transformation is performed in a special gauge field beyond the SM.
- Since these isotopic states can occupy two positions in that gauge field, it proved to have a *spin-like property*.
- Therefore, we called this property as *IFCS*, denoted it by Δ, and we refer to the invariance transformation that we identified – at least by theoretical calculation – as *IFC gauge transformation.* This assumption made the existence of a local gauge field likely, in which the IFCS can rotate and occupy two states. This assumption concluded also a conserved (non-Abelian) current and a corresponding class of $SU(2)$ isomorphic invariances.
- For the "same" object can behave in two ways – for example, in the electromagnetic field, once as the source of a Coulomb force, and in another frame of reference as a source of a (kinetic) magnetic Lorentz force (cf., covariance principle) – they must be able to get transformed into each other. Non-Abelian character and arbitrariness involve that the orientation of the IFCS is of no physical significance. Since we determined the proper form of this invariance transformation (cf., Section 7.3), this gauge transformation counteracts the loss of symmetry between the two kinds of field-charges and brings our equations in compliance with experimental results.
- This invariance shows certain formal similitude to YM-type invariances (Young and Mills, 1954). However, it differs from them, at least, in two features. Once, the concerned physical property, namely the isotopic field-charge (IFC), is a quite different physical property than the *isotopic* states of nucleons. Secondly, the gauge field **D** – and consequently the HySy gauge transformation that rotates the *isotopic field-charge spin* (IFCS, Δ) in this gauge field – is quite different from the isotopic gauge fields introduced for the *isotopic spin* and the *weak isospin* transformations. (For specification, see Sections 7.3–7.5.)

The existence of such an invariance transformation provided us with the (HySy) symmetry, and with the consequent conservation law, namely the conservation of the introduced new property (Δ) of the field-charges. The conservation of IFCS is identical with the requirement of invariance of all interactions under IFCS rotation (in the gauge field **D**, where it is interpreted). This means that all physical interactions should be invariant under a transformation in an isotopic gauge field, more precisely, under a rotation of the property, called IFCS (Δ).

The IFCS (Δ) is attributed to the field-charges as entities, independent of which physical interaction field's charges are considered. Its derivation introduced not only a *property* that differs from all described ones but also a *field* in which it can be interpreted. Concluding the existence of the conserved property called IFCS (Δ) assumed the existence of a gauge field that has been introduced partly analogously to the YM field. However, it differs from that field, for there is a quite different (i.e. from the isospin) property that should be rotated and taking two stable orientations in this (different) gauge field. We found this field (**D**) beyond the SM in Section 7.3.

Table 9.1: Field-charges in the standard model.
All the four field-charges are doubled (split into isotopes)
in the IFC model.

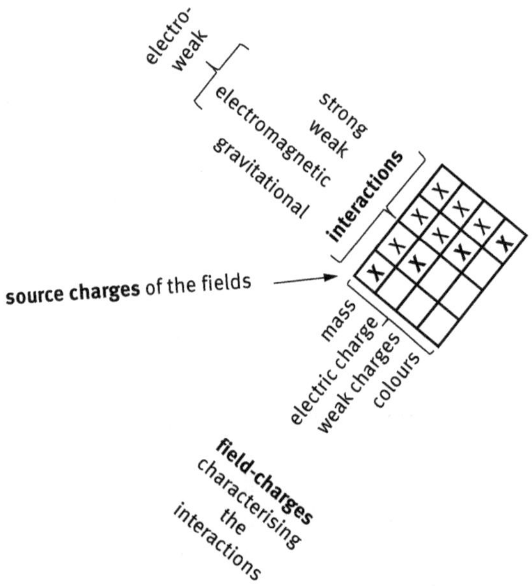

One can read from Table 9.1 that *mass* characterises the particles participating in all the four interactions. At the same time, it is the source of only the gravitational interaction. The gravitational field leaves all the three other field sources intact.

All field sources are characterised by an *electric charge* but the gravitational. The electric charge is the source of only the electromagnetic interaction. The electromagnetic field leaves intact the field sources of only the weak and the strong interactions.

Weak charges characterise the particles participating in the weak interaction. Weak charges of quarks and leptons in all the three families and six flavours are the sources of the weak interaction. The weak charges are defined by the Gell-Mann–Nishijima formula: $Y_W=2(Q-T_3)$, where Y denotes the weak hypercharge, Q the electric charge, and T_3 the third component of the weak isospin. From among the carriers of the weak charges, the quarks participate in both the weak and the strong interaction. However, the leptons destroy the symmetry of the picture, because they do not take part in strong interactions, only in weak. All weak charges leave intact the strong colours.

Colour charges characterise particles that are the actors of strong interaction. Colour is the source of that interaction. No colour charge characterises the sources of the other interactions. The carriers of the strong colours, the quarks, have also mass, electric charge and weak hypercharge.

The boxes in Table 9.1 are left empty where a field-charge does not appear in a given interaction.

Bold crosses mark the field sources of the individual fundamental interactions.

All about Table 9.1 is in accordance with the SM. In the IFC theory, source charges of all the four fields are doubled, what means that they split into isotopic states.

We have not introduced new particles in the theory up to this point. The twins of the fermions (and mass units) were present in our theories for long. They were present in the components of the Hamiltonians (and Lagrangians) as their sources. We only have not paid attention to the difference between the members of these twin pairs. Although the difference between the gravitational and inertial masses was known for over three hundred years, the difference between the Coulomb and Lorentz charges also for much longer than a century – both associated with the bound and free, the potential and kinetic states, as well as with the scalar and the vector components of the field sources – there has not appeared a need to handle them as separate particles. Due to their analogous roles of the sources of the gravitational and electromagnetic fields in the Hamiltonians, we had good reason to assume that the sources of the weak and the strong interaction fields split also according to the bound and free states of their agents. So, all these twin siblings were at our hands. Now, we found a new property that makes it possible to distinguish them. The theoretical prediction of the existence of this property that we called IFCS and denoted with Λ leads us to draw new conclusions.

9.3 Quanta of the D field

When we introduced the distinguished two kinds of IFC in our physical equations we reached a limit: we have got equations where certain symmetries of the traditional equations were distorted. That was not in harmony with our experience (that means, was not in accordance with known facts). Then, we referred to a conservation law

(Darvas, 2009) for a newly introduced quantity, namely the IFCS (Δ), and its invariance is expected to restore (see Section 3.5) the lost symmetry.

Invariance between $\urcorner_V$ and $\urcorner_T$ means that they can substitute for each other arbitrarily in the interaction between field-charges of any given fundamental physical interaction. Invariance means also that particles, disposed of these properties, can be exchanged. The "exchange rate" (gauge) depends on the velocity of the kinetic field-charge compared to the respective matter field (i.e. to the potential field-charge in rest in that field).[9.3] The validity of the assumption can be verified by demonstrating the existence of the gauge bosons that mediate the exchange.

The IFCS should exist in an earlier-defined kinetic gauge field **D**. According to the SM, such a gauge field must have quanta that carry the IFCS Δ. The exchange of the value of Δ mediate the transformation of $\urcorner_V$ into $\urcorner_T$ and back, so that the expected quanta switch the emitter field-charge $\urcorner_V \rightarrow$ to $\urcorner_T$ and the recipient field-charge from $\urcorner_T \rightarrow$ to $\urcorner_V$ and *vice versa*. These quanta have not been observed. Therefore, they do not belong to the class of our observed quantities. A test of the derived theory is whether one can find such mediating bosons. *This is the first issue in this work, where we left the field of reinterpretation of known facts for a* terra incognita *and moved to the domain of prediction.*

We denote the *predicted* quanta of the **D** field by δ. We call this (up to now) hypothetical boson "*dion*", after the ancient Greek term meaning "flee", "flight", "rout" in English. The δ quanta (dions) carry the Δ (IFCS as a physical property: charge of the **D** field).[9.4] The well-known SM mediating bosons carry the quantum numbers between the interacting particles characterised by the $J^{(1)}$ current as we learned in the

9.3 Jackiw and Rebbi (1976) (with assistance by t'Hooft) – investigating the YM pseudoparticle solution by Belavin, Polyakov, Schwartz and Tyupkin (1975), which was found to be $O(5)$ invariant, and after having applied the solution to the Dirac equation – demonstrated that the pseudoparticle was distinguished by possessing a large *kinematical invariance group* "possibly important in future developments of the theory", although they did not analyse these kinematic consequences for the Dirac equation. Now, we have here a YM type pseudo(?) particle invariant under a "kinematic" group.

9.4 The name dion brings about some associations of the *dyons*, which were proposed first by Schwinger (1969). Dyon was presumed as a hypothetical particle endowed with both electric and magnetic charges. From our aspect, it can be considered as the first idea of doubling the properties of a charged physical object. Although that doubling of properties differed from our distinction between isotopic electric charges, it had in Schwinger's interpretation a similar feature: it distinguished the Coulomb-like charges from magnetic charges, which were assumed as results of the velocity of current-like (kinetic) charges. Schwinger introduced this concept when he extended the quantization condition, set up earlier by Dirac, to the dyon. By the dyon model, Schwinger predicted a particle with the properties of the J/ψ meson, a few years before it was discovered in 1974. A sign of the actuality of the topic is a renaissance of the discussion of dyons in the literature of non-Abelian theories (Singh and Tripathi 2013). The latter refers to octonion electrodynamics, which is widely used in the literature, and which we reduced to quaternions in Darvas (2012a, e) (concerning earlier treatments, cf., e.g. 't Hooft (1974) and Julia Zee (1975)).

SM. Dions are bosons predicted beyond the SM – with a notice in the next paragraph. The dions are expected to mediate a non-SM interaction (anti)parallel with the SM bosons governed by the (also conserved) $J^{(2)}$ current (cf., Section 7.4). Therefore, all the quantum numbers of the interacting field-charges that are mediated by the respective SM mediating bosons carry the known SM quantum numbers, and accordingly, the known quantum numbers of a dion (δ) must be 0, with the exception of the quantum number of the respective Δ (what is beyond the SM). Since Δ transforms under the group of HySy, according to the τ algebra and its $SU(2)$ compatible group learned in Section 5, we can agree in a free convention for the sign of the IFCS Δ to be + or – (½) according to whether it switches the IFCS from a potential state to a kinetic or back. Thus δ carries a value of ±1 IFCS Δ between the interacting IFC.

We have to add a notice about the mass and electric charge of the dion. The mass of a mediating boson in the SM depends on the range and strength of the interaction. The masses of the gauge quanta of the individual interacting fields are in accordance with this requirement: they carry that mass. Long-range interaction's bosons (like the graviton and the photon) have 0 mass, as well as the strong gluons (due to the asymptotic freedom). Short-range and symmetry breaking interactions' bosons that mediate the weak interaction have relatively large masses. What can we expect from another boson that exists in a simultaneously present gauge field and acts parallel (or antiparallel) with the known SM exchange bosons? Note that observing from a given reference frame, the field-charges in potential state (m_V, q_V) are *bare quantities*, while the kinetic ones owe "*dressed*" or *renormalized* mass and electric charge density (m_T, q_T) in the same frame. When an IFCS invariance mechanism (cf., Section 10.3) changes the value of Δ to the opposite, the mediating boson must carry the difference there and back, if the observer insists on his chosen reference frame. We have no reason to give up this physicist observer position. Therefore, the dion is supposed to owe this mass and electric charge density difference. The nature and origin of this mass will be discussed in Section 10.3.

Another strange peculiarity of the dion is that it switches $\daleth_V$ and $\daleth_T$ between scalar and vector quantities in their respective fields.

If at first, the idea is not absurd,
then there is no hope for it.
Albert Einstein

Chapter IV: INTERACTION BETWEEN ISOTOPIC FIELD-CHARGES

10 ISOTOPIC FIELD-CHARGES ($\daleth_V$ and $\daleth_T$) IN INTERACTION

The first preliminary assumption of this book, the definition of isotopic field-charges (IFC) and, following from it, the conjecture of the existence of isotopic field-charge spin (IFCS), was discussed in the previous sections.

The second preliminary assumption of this book is to make conjecture interaction between IFC.

Before analysing possible interaction between different – isotopic – states of field-charges of a given field, we refer to the history of how interactions have been treated. That was discussed in Section 3, looking back to the classical years of quantum electrodynamics (QED) (1928–1932).

In short, we had two models in classical QED. One, in which the interaction starts between two bound states of particles, and another, in which the interaction starts between two free state particles. In the first model, the effect of the vector potential enters only in the perturbation, in the second model, the role of the scalar potential is left for the perturbation. Both pictures are approximations, and both lead to relatively good results in accordance with the experience.

10.1 Mechanism of the interaction between IFC

Now, we are introducing an *intermediate model* in which the interaction takes place between a bound and a free state particle initially. Please, note that if the model of the initial interaction between two bound states and the model of the initial interaction between two free states could have been considered equivalent (Bethe and Fermi, 1932), there can be found an equivalent place among the physical descriptions for the intermediate model as well. However, such a model has not been described.

Nevertheless, the intermediate model allows a further interpretation: an exchange between the two isotopic states of a single particle. To understand the character of the interaction between two isotopic states of field-charges better, first, we will investigate this option – an interaction between the isotopic states of a single particle – in the following subsection.

10.1.1 Single particle's IFC states

In order to understand the possible mechanism of the intermediate model of interactions, we discuss the change in isotopic states of field-charges first on the example of single particles.

https://doi.org/10.1515/9783110713183-010

When we take a measure on an object, we have no experience whether we found it in one or the opposite IFC state. When we observe a single particle, it will be either in one or in the other IFCS state. We can call the two states either as potential and kinetic, scalar and vector, or bound and free states. Our measurement's data series record an entanglement (mixture) of the two states. Nevertheless, we do not observe individual IFCS states. Our observation suggests that they behave as being in both states, each measured particle can occupy both a potential (bound) and a kinetic (free) IFCS state. In the lack of experience to catch a particle in one or the other stable state, we have good reason to assume that they permanently change their states. (They do it randomly or with a stable frequency, we do not know, they may probably follow a similar mechanism as quarks do during their colour change via gluon exchange.) For a single particle's behaviour, let us notice the following observations. Provided that the hypersymmetry (HySy) model allows single-particle (self)interaction, the energy of a single particle could oscillate between the potential (scalar) and the kinetic (vector) component of its Hamiltonian. Its full energy may appear once in one and then in the other. Accordingly, the particle can oscillate between its two IFC states. One does not observe this change in its state at low velocities, because it does not involve recordable mass change (i.e. recordable mass-dion exchange).

Let us consider a model of a doublet when a particle can be in a potential state (V) and in a kinetic state (T). According to its actual state, it has either potential or kinetic energy respectively. According to our observation, all particles possess both. We can interpret the phenomenon in the following four ways:

- probabilistic model
- harmonic oscillator model
- flip-flop model
- intermediate particle model

10.1.1.1 The probabilistic model

In this model, we can consider that the wave function of the given particle may be in a potential state with amplitude ψ_V, or in a kinetic state with amplitude ψ_T. The wave function of the given single particle is a mixture (entanglement) of the probabilities being in the potential or the kinetic state:

$$\psi = \begin{pmatrix} \psi_T \\ \psi_V \end{pmatrix}$$

We detect this probabilistic mixture in a measurement. In a large set of particles (e.g. in the case of a massive body consisting of many particles), the probabilities reach a stable proportion and we observe stabilised measurable potential and kinetic energies in a given reference frame.

10.1.1.2 The harmonic oscillator model

This presumes that – in the simplest case and in first approximation – the Hamiltonian of the particle consists of V and T only. The model assumes that the energy of the single particle is concentrated either fully in V or fully in T and the energy commutes periodically between the two states. The transition is assumed to follow a harmonic function in time. We can take a measurement of the energies in a transitional phase, which can be interpreted as a mixture of probabilities being in the two extreme states, similar to the wave function in the probabilistic model. However, in the case of a single particle, physical meaning can be attributed only to the two extreme states. As mentioned, in this model the energy of the particle is concentrated either in V or in T. This has certain consequences.

When our physical object (i.e. field charge, denoted by $\daleth$) is in the *potential state* IFC_V – or in our notation $\daleth_V$ – (*its full energy is V*), its features show up as a *scalar particle* and it behaves as a *source of a scalar field*. When it is in the *kinetic state* IFC_T – or in our notation $\daleth_T$ – (*its full energy is T*), it behaves as a *wave*, its field-charge current takes the form of a *vector*, and *generates a vector field* around itself.

Unlike the field-charge of the scalar potential field, whose quantity is invariant (bare or rest field-charge), the quantity of the field-charge in kinetic state (more precisely, the three independent components of its current) varies according to its velocity in the respective reference frame where it is actually observed (according to the Lorentz transformation and in accordance with the general covariance principle).

The harmonic oscillator model presumes the permanent change of a single particle between its two IFC states along with the changes in the listed consequences. Although one cannot observe the differences between the two IFC states at low velocities (energies) – due to the absence of recordable mass difference, when one detects a particle in a given time moment, one can distinguish whether it appears as a corpuscle or a wave.

10.1.1.3 The flip-flop model

This model differs from the harmonic oscillator model that the change between the two IFC states is assumed to undergo suddenly, without transition. This model considers that the IFCS (Δ) can rotate in a presumed **D** field not smoothly, rather by a sudden switch and it can take only two quantum positions (as discussed in mathematical terms in Section 7.4).

10.1.1.4 The intermediate particle model

This model presumes that the switch between the two states happens with the mediation of a δ boson that governs the transition from one IFC state to the other. This boson may not carry any physical quantum number but the Δ. Therefore, it cannot be identical with any existing known (standard model (SM)) intermediate boson. It should act parallel (or anti-parallel) with them. This is why we assume it as the (two eggs) twin brother of the SM bosons.

The common feature in the four models is that any single particle carrying a field-charge of a SM fundamental physical interaction field is in permanent change between its two IFC states. Since it is difficult to observe them in one or the other IFC state we attribute both properties to each of them. This is the background of the long-rooted *corpuscle-wave double behaviour* discussion.

Free particles can be observed only in idealised, extreme physical situations. Provided we can detect a particle really in a free state, it must be in a kinetic state since in the absence of another particle which it could exert an effect on, there is no reason to presume its scalar potential field. However, in measurement situation (detection), no particle without interaction (i.e. in a real continuous free state) can be imagined. We will show later, how one can conclude the kinetic state of a particle in a bound state.

Bound particles oscillate permanently between the two states so that the potential and kinetic IFC states of the particle commute between each other. The assumed mechanism will be characterised in Section 11.1.

As mentioned, a particle *in a potential state* behaves like a *corpuscle*, while *in a kinetic state* it behaves like a *wave*. Since the two IFC states switch between each other with a high oscillation frequency one cannot detect a single particle in a full corpuscle or in a full wave state. What we can detect depends on the nature of the actual interaction of the individual particle on the one hand, and we can observe a probabilistic mixture of the two states on the other hand.

A particle *in a potential state* plays the role of the *source of a scalar field*. Therefore, a potential IFC (we denoted it by $\daleth_V$) is a scalar quantity. A particle *in a kinetic state* serves as a *current component source of a vector field*. So, a kinetic IFC (we denoted it by $\daleth_T$) appears in the role of a coefficient assigned to three vector components according to three, directed, independent components of a field-charge current. An important consequence of the switch between the two IFC states is that the IFC must commute between a scalar and three components of a vector quantity (as described in the algebra of HySy, cf., Section 5).

10.1.2 The intermediate model of interaction between two particles

We cannot reconstruct whether Møller (1931) was aware of the same mechanism in 1931 that we are proposing now. Probably, he was not. Nevertheless, his original formulation, which he derived, contained a similar asymmetry between the roles of the interacting particles which we are proposing. Bethe and Fermi (1932) considered this asymmetry an incompleteness of the theory by Møller and rectified it. Instead of their artificial correction in the scattering matrix elements, we attempt to interpret the original equation and fit it in an alternative picture taking into consideration another version of the same facts.

In the intermediate model, the interaction starts initially between a bound and a free state particle. In our terminology, it takes place between a potential IFC state ($\urcorner_V$) and a kinetic IFC state ($\urcorner_T$) particle. This model corresponds to the requirement of the validity of the Pauli principle for IFC discussed in Section 3.

This intermediate model can be interpreted in different ways. Remaining at the example of QED, we can interpret that the Coulomb potential of a particle in an initial state interacts with the vector potential of another. Another possible interpretation is more general: the potential (scalar) part of a Hamiltonian interacts with the kinetic (vector) part. We can also say that the bound state of a particle interacts with the free state of another particle. (This statement sounds strange in this form since as soon as the latter particle gets subject of interaction, it will be no longer free. At least, at the moment of interaction, they form a quantum system, where they cannot occupy the same IFC state.) Finally, using the terminology introduced in the IFC theory, the interaction can take place between particles in two opposite IFC states. The latter three interpretations do not reduce their relevance to the electromagnetic interaction.

All these three models (the Dirac–Breit–Fermi model in accordance with the Heisenberg–Pauli formalism on the one side, the Møller–Bethe–Fermi model on the other, and finally the intermediate model based on Møller's presumed original intention) are interpreted for the interaction between two particles.

Returning to the transformation of mass, let us investigate again the example of a dropped and rebound ball (cf., Section 2). We cannot observe a single elementary mass in itself, because, at the moment, we have no instruments to observe objects at the Planck scale. If we would drop such a single elementary mass, its gravitational mass would change into inertial mass during the fall, then, when rebounding, transforms back into gravitational mass at its upper rest position. Since a real ball consists of many elementary masses, like the Earth that attracts it, there appear both gravitational and inertial masses in it in a statistical distribution. This characteristics presents itself in the intermediary states during fall when it moves, on the one hand, and is the subject of attractive force as well, on the other hand, simultaneously. What would happen, if it consisted of a single elementary mass charge? This elementary mass would exchange states with the mass with which it is in interaction (i.e. with one of the elementary masses of the Earth attracting it). At a moment, one of them was in the state of gravitational and the other in the state of mixture of kinetic and gravitational mass points; then at a next moment, in the opposite, the elementary masses of the Earth were in mixture of gravitational and kinetic states and the dropped single elementary mass in a kinetic state. In case of a real ball, a part of its constituting elementary masses is in the gravitational state, another part of them is in the inertial state, they all commute between the two states, and the proportion of their number within the ball changes when the ball falls. There must be something that mediates this commutation.

Summarising the proposed mechanism of the interaction between two particles: (1) first, we assume that a single particle can oscillate between two IFC states – as we saw the possible mechanisms in the previous subsection – that means, they are at any moment either in a potential or in a kinetic state; (2) second, we assume that interaction takes place always between a particle in one of the IFC states and another particle in the opposite IFC state (cf., Pauli principle); and (3) third, the two particles switch their IFC states simultaneously during the interaction so that both remain in an opposite IFC state at a given moment.

All the described interaction mechanisms were founded mathematically in Sections 7.2–7.4. We will present the mechanism of the IFC interaction in the individual physical interactions in Sections 11.2–11.5.

10.1.2.1 Symmetric or asymmetric interacting agents?

The problems of the consideration of asymmetries between interacting particles and antisymmetrisation as a method were not far from Bethe when he proposed to Fermi to symmetrise Møller's scattering matrix in course of studying transitions of interacting electrons between two states in 1932. The method appears in his *Ansatz*, written in the previous year during his first visit to Rome, in which he already expressed his thanks to Fermi for his advises. (The idea of reduction of dimensions, in another context, appears a few years later also in Wigner's little group.) The Bethe *Ansatz* concerned the interaction of two electrons being in opposite spin positions that respect the Pauli exclusion principle. Seemingly, Bethe and Fermi did not accept similar asymmetric roles (of another property) for the interacting electrons when applying the Møller scattering model, and they insisted then on symmetry.

The symmetric interchangeability of the interacting fields used to be a strong paradigm. It prevailed physical thinking for long, even in recent decades.[10.1] Not only Bethe and Fermi (1932) insisted on the symmetry in the interacting physical agents. Note that much later, Goldstone (1961) assumes also a symmetric role between the interacting scalar fields. P. Higgs (1964) 'breaks' this symmetry in the roles of the interacting scalar fields using a seemingly similar consideration like Møller (1931). Our approach differs from Higgs' in the assumption of velocity dependence in the case

10.1 The unquestionable faith in the "perfection" (i.e. symmetry) of the world that started in European way of thinking by Herodotus, namely what meant that when theoretical considerations (suggesting a more perfect, that is, symmetric model) and empirical experience (demonstrating symmetry violation) conflicted, scholars bet on the former, and this tradition was broken first by Kepler. For a thinker, from ancient Greece up to the end of the Renaissance, faith in the symmetry of the world was stronger than what was actually experienced. When a decision had to be made between faith in symmetry and empirical experience, the advantage was given the former. This is how symmetry was able to become a principle (Darvas, 2007).

of the vector field. S. Weinberg (1967) attributes the asymmetry between interacting agents to chiral configurations. In our "covariant interpretation", the interacting agents must be in different physical states, but their roles must be exchangeable. The latter can be interpreted in terms like Yang and Mills formulated, only we extend their terms of "arbitrary choice" of one or the other state of the interacting particles in an *isotopic field* → to the *IFC field.*

What we are doing here is to apply the idea of asymmetry to the interaction of particles that must be in opposite IFC states, so that we acknowledge the asymmetric roles of the agents entering into interaction with each other. In analogy to the Pauli principle formulated for spin, we assume that particles (not only electrons and not only electrically charged ones) should be in opposite IFC states when they interact; in other words, they must owe opposite position IFCS. Since IFCS is interpreted in a gauge field (what we denoted by **D**[10.2]), and it concerns field-charges of any field, we have some reason to conjecture that the assumed principle of interaction between opposite IFCS particles can be extended to any interaction field. Thus, IFCS may play an integrating role among the fundamental physical interactions. The optimism is cherished by the successful extension of the Bethe *Ansatz* to several areas of quantum field theories, covering all physical interactions respectively, at an accelerated pace in the recent two decades.

10.2 Interpretation of the IFCS conservation

Invariance between $\urcorner_V$ and $\urcorner_T$ means that they can substitute for each other arbitrarily in the interaction between field-charges of any given fundamental physical interaction. (The possible mechanism was discussed in the previous Section 10.1.) They appear at a probability between [0, 1] in a mixture of states in the wave function:

$$\psi = \begin{pmatrix} \psi_T \\ \psi_V \end{pmatrix}$$

so that the energy of a *single particle* oscillates between the *V* and *T* components of the Hamiltonian, while the Hamiltonian of a *composite system* is an entanglement (mixture) of the oscillating components characterising the energies of particles that constitute the system. The individual particles in a *two-particle system* are either in the *V* or in the *T* state, respectively, and switch between the two roles permanently; while the observable value of *H* is the expected value of the mixture of the actual states of the two, always opposite state particles.

10.2 Note that the electromagnetic field was denoted by **A**, the YM isotopic field by **B**, and C was occupied by several other purposes in physics, among others in the name of the theory of strong interactions: QCD for **C**olours.

The invariance between $\daleth_V$ and $\daleth_T$ (what is guaranteed by the conservation of Δ), and their ability to swap means also that they can restore the symmetry in the physical equations, which was lost when we replaced the general $\daleth$ (namely mass m, electric charge q, etc.) by their isotopes $\daleth_V$ and $\daleth_T$.[10.3]

10.2.1 On the roles of the masses once again

Probably, the most remarkable issue for our discussion is the exceptional role of the mass: it takes part in all interactions (cf., Table 9.1). We have a look at this exceptionality in the following. That will give the inspiration to admit further evidence for the two kinds of IFC.

The two IFC states of an object are indistinguishable in an everyday experiment. First, because – as we will saw (cf., Section 11.1) – a single particle changes its IFC state permanently; second, it is difficult to observe a *free* single particle; third – as we showed also (in Section 10.1) – particles in interaction exchange their IFC states between each other at a high frequency; and finally, when we observe them, we can measure a mix of their two IFC states.

In general, we observe objects composed of a set of particles, whose states within the single object are also a mixture of the two states. So we do not have macroscopic experience on the IFC states. The total energy $H = T + V$ (we are neglecting here possible further components) of a complex object reflects an average of the particles individually in states ψ_T and ψ_V – at least according to the model described in this book.

Nevertheless, there are phenomena that allow us to assume the existence of the individual IFC states. Let us investigate a few simple examples.

First, look at the motion of two bodies with equal rest masses $m_0 = m_1 = m_2$. They revolve in their gravitational field around their common centre of mass, which is in the half point of the straight line connecting their individual centres of masses. They will move like we rotated a barbell around the centre of its connecting bar. In the isotopic field-charge model, we can imagine their interaction so that the gravitational mass of body$_1$ m_{V1} exerts a force on the inertial mass of the body$_2$ and forces it to move in a 'central' (sic!) scalar force field, making the mass of inertia m_{T2} to revolve around their common centre of mass. Similarly, the gravitational mass of body$_2$ exerts a force on the inertial mass of the body$_1$ causing a similar movement around their common centre of mass. We are unable to distinguish the two isotopic states of the bodies because once, nearly 50–50% of the elementary particles in both bodies are in the opposite isotopic mass states, and second, all change this state permanently, thus

10.3 Consequences of the application of effective field theories were analysed, for example, by E. Castellani (2001) in philosophy and by S. Weinberg (1997) in physics.

one cannot identify which is in gravitational and which is in inertial state at a given moment. However, we can idealise the classical situation by a purely theoretical assumption that in a certain moment all constituting particles of $body_1$ are in state V, and those of $body_2$ in state T (then they change). We can quantify the forces between the two bodies in the following way (third law of Newton):

$$\mathbf{F}_{1\to2} = f\frac{\mathbf{r}\cdot m_{V1}}{|r|^3}m_{T2} \quad \text{and then} \quad \mathbf{F}_{2\to1} = f\frac{\mathbf{r}\cdot m_{V2}}{|r|^3}m_{T1}$$

where f is the gravitational coupling constant. This means that either the gravitational mass of the $body_1$ exerts a force on the inertial mass of the $body_2$ at a moment or *vice versa*. This process switches repeatedly at an unobservable pace. At both sides of the interaction, both masses behave once as a mass of gravity and then as a mass of inertia, depending on whether they serve incidentally as a source of a force, or as they suffer the acceleration caused by an external force. We did not take into consideration any general relativistic effect in this idealised model.

Now, let us investigate a celestial system, consisting of two objects, with rest masses m_1 and m_2. Let it be $m_1 \gg m_2$, thus their common centre of mass is near to the centre of mass of the m_1, inside the $object_1$. The m_{V1} mass of gravity of the $object_1$ will exert a force on the $object_2$. The m_{T2} mass of inertia will suffer a force $\mathbf{F}_{1\to2}$. This force will accelerate it, and this will cause the $object_2$ to revolve around its common centre of mass (around $object_1$). Note that the interaction took place between the mass of gravity of the $object_1$ and the mass of inertia of the $object_2$. For the sake of symmetry, there is another (although much less demonstrative) interaction between the mass of gravity of $object_2$ and the mass of inertia of $object_1$. Nevertheless, in this case, we could neglect the mass of gravity (as a property) of the $object_2$ and the mass of inertia (as a property) of the $object_1$.

Third, let us investigate two objects with different quantity masses ($m_1 \gg m_2$), and with opposite sign, but equal in absolute value electric charges (q, $-q$). Let us assume – for simplicity – that they are point-like and they save this property during an interaction. The classical Bohr model of an H atom was an idealised model for such a system. The Bohr model could not be applied in quantum mechanics for two reasons, once for the quantised energies and momenta of the electron around the proton, second, for the electron could not be imagined point-like for its wave function's runaway. The quantised orbits of the electron do not influence our example. Regarding the runaway of the wave function, we must advance one of the consequences of the IFCS model. We proved the possibility of the described mechanism of the permanent interchange between the IFC states (cf., Section 10.1). According to that, there is no time left for an electron in a free state to run away; it switches its IFC state more frequently than it could reach a too large extension before next switch. So it can be considered quasi-point-like. More precisely, when it is accidentally in a kinetic state, it 'feels' itself asymptotically free and behaves like a wave, so it tries to run away. It has not too much time to spread out significantly because as soon as it switches accidentally to a potential state, it 'feels' itself bounded and behaves like a corpuscle. (Cf., the harmonic oscillator model in Section 10.1.1.2.)

It is easy to imagine how the two objects would interact in an idealised quasi-classical H model: they will revolve around each other. How does it happen? There emerge two forces between them, a gravitational and an electric. The former – being magnitudes weaker – can be disregarded. That is, their *mass of gravity* does not play a role in their interaction and mutual motion. The electric (Coulomb) force originates from their electric charges. This force is equally strong for both the (q) and the $(-q)$ charges (irrespective of the mass difference between them). If this force would affect the electric charges, they would revolve around their geometric centre (= the centre of charge, the half point of the straight line, connecting them), as we saw on the equal mass example of the barbell-like model. They do not. The Coulomb charge of the object_1 exerts a larger acceleration for the less massive $(-q)$ object_2, and the same absolute value Coulomb charge of the object_2 exerts a smaller acceleration on the much more massive (q) object_1:

$$\mathbf{F}_{1\to2}:\ k\frac{\mathbf{r}\cdot q_{V1}}{|r|^3}(-q_{T2}) = m_{T2}\cdot\mathbf{a}_2$$

and

$$\mathbf{F}_{2\to1}:\ k\frac{\mathbf{r}\cdot(-q_{V2})}{|r|^3}q_{T1,} = m_{T1}\cdot\mathbf{a}_1$$

where k is the electric coupling constant. Since the left sides are equal and $m_{T1} \gg m_{T2}$, $\mathbf{a}_2 \gg \mathbf{a}_1$.

The two objects revolve around their mutual centre of mass, which is close to the centre of the (m_1, q), and do not around their mutual centre of charge. That is, the Coulomb force causes acceleration reciprocally proportional to their inertia. This inertia is characterised and quantised by the *masses of inertia* of the $(-q)$ and the (q) charges, respectively. We did not take into consideration any kinetic Lorentz force and relativistic effects in this model (unlike we did in Section 3.5).

On the example of the described idealised (classical) model, we observe a phenomenon, where we *excluded the role of the mass of gravity*, but we *observe an effect proportional to the mass of inertia*. This is another *argument to assume* that *the masses of gravity and inertia* are *two different properties* of the agents in the system. Mass plays a role in the effect of an electromagnetic interaction (and so does it in weak and strong interactions as well), which underlies its unique role. At the same time, this latter was an example of an effect where only the mass of inertia took part, the mass of gravity did not.

10.3 Mass of dions that mediate HySy transformations

We saw in Section 9.2 that mass is the only field-charge that is present in all the four fundamental interactions (cf., Table 9.1). We demonstrated in Section 10.2.1 that all kinds of forces in the individual interactions affect the mass of the field source, instead of the characteristic field-charge of the given field. Mass plays an odd role among field-charges.

We saw also that the potential members of the IFC sibling pairs have rest mass of the respective field source. At the same time, the mass of the kinetic members of the same sibling pair is their dressed (Lorentz boosted) mass. The dions predicted in Section 9.3 should convey the difference between those two mass values during transformation from the former IFC state to the other, and back.

As we mentioned already, theories are generally considered relativistic if they meet the condition to be invariant under Lorentz transformation. For example, GTR and QED meet this condition. However, derivations of all the Einstein equations and their solutions, as well as the Dirac equation (and related other discussions of QED) are performed with assuming preliminary approximations to "not too high velocities". In fact, invariance under the Lorentz transformation is a necessary condition for a theory to be relativistic, but that condition is not (always) sufficient (Darvas 2015, 2018b). High-energy experiments reach velocities very near to that of the light. In the interpretation of their results, one can disregard effects of those high velocities no more, as we saw in Section 6. The absence of such precise high-velocity considerations led (among others) to anomalies that formulated the demand for extending the SM in the 1990s. One of those alternatives is the subject of this book. We introduced (cf., Section 6.2) a velocity-dependent field (**D**), which proved to be a gauge field, and should be added to the SM fields. Nevertheless, the field **D** is interpreted beyond the SM. The intermediate bosons of the **D** field are assumed to be the dions that mediate the IFCS between IFC-s.

However, a problem showed up by itself there. According to gauge theories, interaction mediating, spin-0 bosons must be massless. The theory of HySy predicted spin-0, but massive intermediate bosons (δ). The mass of intermediate bosons in spin-0 fields must arise from dynamical symmetry breaking (Goldstone, 1961; Glashow, 1961; Goldstone, Salam and Weinberg 1962; Weinberg 1967). The mass of δ should arise from spontaneous breaking of the group of HySy. The group of HySy has two free parameters. Its spontaneous breakdown may eliminate one of them: it allows to perform a transformation that does not influence the physical state of the investigated system. The other free parameter can be discussed in terms of the BEH mechanism (named after the discoverers Higgs, 1964; Englert and Brout, 1964). The intermediate bosons of the SM belong to one of the three Goldstone boson types defined by Weinberg.[10.4] The

10.4 Originally, it was assumed that the spontaneous symmetry breakdown responsible for the intermediate vector boson masses was only due to the vacuum expectation values of a set of spin-0

simultaneously appearing HySy δ boson belongs to the fictitious Goldstone bosons, whose mass is removed by the Higgs mechanism. The mass of the fictitious Goldstone bosons is eliminated by the unitarity gauge condition. (HySy meets that condition.) According to the simultaneous presence of a SM interaction's symmetry group and the HySy, their bosons should be rotated simultaneously ($G_{SM} \otimes G_D$).[10.5]

In the course of interactions between two fermions, the bosons of the **D** field are expected to be exchanged (anti/)parallel with the exchange of a SM boson (Darvas 2011), since the **D** field appears always as an extension to a SM interaction field. The HySy theory must try to avoid affecting the respective SM bosons. For this reason, when the HySy theory assumes that the dions obtain their mass by a transformation of the non-SM **D** field via a BEH mechanism, it should guarantee that the respective SM boson be left intact. The latter condition demands the existence of a transformation matrix that includes a particular transformation angle (like the fermion flavour mixing θ_C Cabibbo angle). This rotation – that can be characterised by an inclination angle θ_D – can guarantee that an expected 0 mass (Goldstone) boson would transform into the predicted mass HySy δ boson (or back), while the (anti/)parallel SM boson does not change its properties, similar to the mechanism of the electroweak theory's weak (Weinberg) mixing angle.[5] The existence of such a HySy field rotation mechanism is discussed below.

10.3.1 Mass of the δ boson

The interaction between two isotopes of any field-charge is mediated by a massive non-SM gauge boson (Darvas, 2009, 2011). According to the IFC theory, the mass of this boson (called dion, δ) is the difference between the boosted (dressed) and the rest (bare or invariant) masses. The mass of this gauge boson is independent of the type (gravitational, electromagnetic, weak, and strong) of the interaction: $m(\delta) = m_T - m_V = (\kappa - 1)m_V$ where $\kappa = 1/\sqrt{1-(v/c)^2}$, m_T is the Lorentz boosted mass and m_V denotes the mass that appears in the potential (scalar) part of the Hamiltonian, which is equal to the rest mass of the concerned field-charge.

fields. Later this approach became more sophisticated, and it was assumed that the considered symmetry breaking was of a purely dynamical nature. Weinberg (1976) distinguished three types of Goldstone bosons (fictitious, true, and pseudo-), and, accordingly, three dynamical symmetry breaking mechanisms. We remark that Weinberg's classification allowed the existence of other (at that time unknown) gauge fields and intermediate bosons, which encouraged the elaboration of the HySy field rotation mechanism discussed here. In respect of the latter, cf., also the remarks Sec. IV (3)(B) in Goldstone, Salam and Weinberg (1962).

10.5 Comparison of the rotation of the δ in combination with a respective SM boson, and the mixing of the also massive neutral weak vector boson with γ needs further investigation. Note that the latter are simultaneously rotated (by θ_W) in field **B**, while the former are rotated in field **D** (characterised indirectly by θ_D), as we will see in the following.

The presence of a massive mediating boson assumes a spontaneous symmetry breaking. Spontaneous symmetry breaking rotates the massless Goldston boson plane, producing as a result the massive δ boson and the respective SM mass bosons (here denoted by ξ). In the opposite direction, the same rotation transforms (in its gauge field) the massive HySy δ boson's and the respective SM ξ boson's plane into a massless Goldston boson (δ') while leaving the SM ξ boson intact [cf., eqs. (10.11)+]. In the instance of the isotopic field-charge field (**D**) the quanta of this field (δ) are associated with the conservation of the IFCS (Δ) introduced in physical terms first in (Darvas, 2011). The (inverse) transformation that eliminates 'unwanted' masses produced by the spontaneous symmetry breaking is expected to depend on the velocity of the interacting IFC relative to each other, and assumes the presence of a velocity-dependent (kinetic) *field* (**D**) instead of a simple configuration *space*.

Sections 10.3.2–10.3.3 define the transformation of the fields, whose quanta are the δ bosons. Note that the δ bosons never appear alone. They act simultaneously (parallel or anti-parallel) with one of the SM bosons. Therefore, one requires the transformation of the **D** field together with one of the SM fields (denoted here by $\boldsymbol{X}_{SM}$). Note also that the derivation of the field equations of the interactions and their solutions included approximations (cf., introduction to Section 10.3). Although all field theories required invariance under the Lorentz transformation, they included restrictions to 'not too high' velocities meaning that those approximations cannot be applied at the high kinetic energies (and the respective high velocities) for the interpretation of data collected in experiments producing large accelerations.

10.3.2 Transformation in a coupled SM field and the D field – The origin of the mass of δ

Conservation of the Δ quantity first requires *invariance under* HySy (cf., Darvas, 2018b). At the same time, the interaction between two particles requires *invariance under the Lorentz transformation.* As it was mentioned several times, invariance of physical theories under the Lorentz transformation is a necessary condition, but it is not always sufficient. In certain instances, the transformation needs to be complemented with others.

A general form of the Lorentz transformation's matrix can be written as follows:

$$\Lambda = \begin{bmatrix} 1+(\kappa-1)\frac{v_1^2}{v^2} & (\kappa-1)\frac{v_1 v_2}{v^2} & (\kappa-1)\frac{v_1 v_3}{v^2} & i\kappa\frac{v_1}{v}\frac{v}{c} \\ (\kappa-1)\frac{v_2 v_1}{v^2} & 1+(\kappa-1)\frac{v_2^2}{v^2} & (\kappa-1)\frac{v_2 v_3}{v^2} & i\kappa\frac{v_2}{v}\frac{v}{c} \\ (\kappa-1)\frac{v_3 v_1}{v^2} & (\kappa-1)\frac{v_3 v_2}{v^2} & 1+(\kappa-1)\frac{v_3^2}{v^2} & i\kappa\frac{v_3}{v}\frac{v}{c} \\ -i\kappa\frac{v_1}{v}\frac{v}{c} & -i\kappa\frac{v_2}{v}\frac{v}{c} & -i\kappa\frac{v_3}{v}\frac{v}{c} & \kappa \end{bmatrix} \tag{10.1}$$

where $\boldsymbol{v}$ is the velocity of the interacting particles relative to each other; v_i are the components of $\boldsymbol{v}$; and $v_1^2+v_2^2+v_3^2=v^2$. This formula holds when the origin of the reference frame is fixed to one of the interacting field-charges and restricted to the situation when the velocity vector arrows from one of the field-charges towards the other (at least, while the velocity is not too high, as we will demonstrate it following eq. (10.18)). At this stage, we do not require any prescription for the direction of the co-ordinate axes. First, we interpret the velocities that define the Lorentz transformation in the configuration space, transformed into the above mentioned velocity-dependent field.

Let us introduce the following notations:

$$\frac{v}{c}=\sin\vartheta;\ \kappa=\frac{1}{\cos\vartheta};\ (\kappa-1)=\frac{1-\cos\vartheta}{\cos\vartheta};\quad \text{and}\quad u_i=\frac{v_i}{v}\,(i=1,2,3)$$

The u_i are unitary length ($u_1^2+u_2^2+u_3^2=1$) vector components, pointing in the direction of the axes of the co-ordinate system. So, the Lorentz transformation can be rewritten in the following forms:

$$\Lambda=\begin{bmatrix} 1+\frac{1-\cos\vartheta}{\cos\vartheta}u_1^2 & \frac{1-\cos\vartheta}{\cos\vartheta}u_1u_2 & \frac{1-\cos\vartheta}{\cos\vartheta}u_1u_3 & iu_1\mathrm{tg}\vartheta \\ \frac{1-\cos\vartheta}{\cos\vartheta}u_2u_1 & 1+\frac{1-\cos\vartheta}{\cos\vartheta}u_2^2 & \frac{1-\cos\vartheta}{\cos\vartheta}u_2u_3 & iu_2\mathrm{tg}\vartheta \\ \frac{1-\cos\vartheta}{\cos\vartheta}u_3u_1 & \frac{1-\cos\vartheta}{\cos\vartheta}u_3u_2 & 1+\frac{1-\cos\vartheta}{\cos\vartheta}u_3^2 & iu_3\mathrm{tg}\vartheta \\ -iu_1\mathrm{tg}\vartheta & -iu_2\mathrm{tg}\vartheta & -iu_3\mathrm{tg}\vartheta & \frac{1}{\cos\vartheta} \end{bmatrix} \tag{10.2}$$

$$\Lambda=\kappa\begin{bmatrix} \cos\vartheta+(1-\cos\vartheta)u_1^2 & (1-\cos\vartheta)u_1u_2 & (1-\cos\vartheta)u_1u_3 & iu_1\sin\vartheta \\ (1-\cos\vartheta)u_2u_1 & \cos\vartheta+(1-\cos\vartheta)u_2^2 & (1-\cos\vartheta)u_2u_3 & iu_2\sin\vartheta \\ (1-\cos\vartheta)u_3u_1 & (1-\cos\vartheta)u_3u_2 & \cos\vartheta+(1-\cos\vartheta)u_3^2 & iu_3\sin\vartheta \\ -iu_1\sin\vartheta & -iu_2\sin\vartheta & -iu_3\sin\vartheta & 1 \end{bmatrix} \tag{10.3}$$

Let us take a general ϑ angle rotation matrix in a three-dimensional space stretched by unit axis vectors u_i:

$$R=\begin{bmatrix} \cos\vartheta+(1-\cos\vartheta)u_1^2 & (1-\cos\vartheta)u_1u_2-u_3\sin\vartheta & (1-\cos\vartheta)u_1u_3+u_2\sin\vartheta & 0 \\ (1-\cos\vartheta)u_2u_1+u_3\sin\vartheta & \cos\vartheta+(1-\cos\vartheta)u_2^2 & (1-\cos\vartheta)u_2u_3-u_1\sin\vartheta & 0 \\ (1-\cos\vartheta)u_3u_1-u_2\sin\vartheta & (1-\cos\vartheta)u_3u_2+u_1\sin\vartheta & \cos\vartheta+(1-\cos\vartheta)u_3^2 & 0 \\ 0 & 0 & 0 & 1 \end{bmatrix} \tag{10.4}$$

However, one can interpret the R transformation of the velocity vector components as they were also projected in the velocity-dependent field. Note that while ϑ is a symbolic notation in the Lorentz transformation, it denotes a real rotation angle in R.[10.6] Let us compare this transformation matrix R with the formula for the Lorentz transformation. In this order, let us decompose R to the following two matrices:

$$R = \begin{bmatrix} \cos\vartheta + (1-\cos\vartheta)u_1^2 & (1-\cos\vartheta)u_1u_2 & (1-\cos\vartheta)u_1u_3 & 0 \\ (1-\cos\vartheta)u_2u_1 & \cos\vartheta + (1-\cos\vartheta)u_2^2 & (1-\cos\vartheta)u_2u_3 & 0 \\ (1-\cos\vartheta)u_3u_1 & (1-\cos\vartheta)u_3u_2 & \cos\vartheta + (1-\cos\vartheta)u_3^2 & 0 \\ 0 & 0 & 0 & 1 \end{bmatrix} + \begin{bmatrix} 0 & -u_3\sin\vartheta & u_2\sin\vartheta & 0 \\ u_3\sin\vartheta & 0 & -u_1\sin\vartheta & 0 \\ -u_2\sin\vartheta & u_1\sin\vartheta & 0 & 0 \\ 0 & 0 & 0 & 0 \end{bmatrix} \quad (10.5)$$

and in a bit extended form:

$$R = \begin{bmatrix} \cos\vartheta + (1-\cos\vartheta)u_1^2 & (1-\cos\vartheta)u_1u_2 & (1-\cos\vartheta)u_1u_3 & iu_1\sin\vartheta \\ (1-\cos\vartheta)u_2u_1 & \cos\vartheta + (1-\cos\vartheta)u_2^2 & (1-\cos\vartheta)u_2u_3 & iu_2\sin\vartheta \\ (1-\cos\vartheta)u_3u_1 & (1-\cos\vartheta)u_3u_2 & \cos\vartheta + (1-\cos\vartheta)u_3^2 & iu_3\sin\vartheta \\ -iu_1\sin\vartheta & -iu_2\sin\vartheta & -iu_3\sin\vartheta & 1 \end{bmatrix} + \begin{bmatrix} 0 & -u_3\sin\vartheta & u_2\sin\vartheta & -iu_1\sin\vartheta \\ u_3\sin\vartheta & 0 & -u_1\sin\vartheta & -iu_2\sin\vartheta \\ -u_2\sin\vartheta & u_1\sin\vartheta & 0 & -iu_3\sin\vartheta \\ iu_1\sin\vartheta & iu_2\sin\vartheta & iu_3\sin\vartheta & 0 \end{bmatrix} \quad (10.6)$$

Now, we see that

10.6 The method to be applied shows certain partial similarity to the derivation of Wigner–Thomas rotation, which applies that a boost (here by $\boldsymbol{v}$) and a rotation are equivalent with the combination of two coupled boosts. In the inverse, they correspond to transformations whose combination produces a (Thomas-) precession. As we will show, one of the coupled boost vectors will precess around the other's arrow, or *vice versa*. However, while the Wigner–Thomas rotation is interpreted in the configuration space, we apply it to transformations projected in abstract gauge fields.

$$R=\Lambda\cos\vartheta+\begin{bmatrix}0 & -u_3 & u_2 & -iu_1\\ u_3 & 0 & -u_1 & -iu_2\\ -u_2 & u_1 & 0 & -iu_3\\ iu_1 & iu_2 & iu_3 & 0\end{bmatrix}\sin\vartheta=\frac{\Lambda}{\kappa}+\frac{1}{c}\begin{bmatrix}0 & -v_3 & v_2 & -iv_1\\ v_3 & 0 & -v_1 & -iv_2\\ -v_2 & v_1 & 0 & -iv_3\\ iv_1 & iv_2 & iv_3 & 0\end{bmatrix} \tag{10.7}$$

or inverted

$$\Lambda=\frac{R}{\cos\vartheta}-\begin{bmatrix}0 & -u_3 & u_2 & -iu_1\\ u_3 & 0 & -u_1 & -iu_2\\ -u_2 & u_1 & 0 & -iu_3\\ iu_1 & iu_2 & iu_3 & 0\end{bmatrix}\mathrm{tg}\vartheta \tag{10.8}$$

The matrices in the second terms in (10.7) and (10.8) are also rotation-like. They suggest precession of the axis defined by the vector $\boldsymbol{v}$ around the u_i velocity components. Thus, the Lorentz transformation (10.8) is expressed in terms of a real rotation minus a precession, and both are functions of the relative velocity of the interacting agents. In the instance of low velocity, the transformation R turns into the traditionally known Lorentz transformation. In the presence of a velocity-dependent field, the transformation Λ must be extended according to the rule expressed in (10.8). The unitary velocity components in the precession matrix can be interpreted also like spatial projections of the velocity-dependent IFCS vectors from a velocity-dependent *field* (**D**) to the configuration *space*.

We recall again that the IFC theory requires invariance under the combination of the Lorentz transformation and the HySy of the IFC transformation. One can check in Section 5.2 that the HySy can be represented by the so-called tau (τ) algebra that (in τ_3 representation) led to two transformation matrices: E and τ_3. Remember:

$$E=\begin{bmatrix}1 & 0 & 0 & 0\\ 1 & 0 & 0 & 0\\ 1 & 0 & 0 & 0\\ 0 & 0 & 0 & 1\end{bmatrix}=\begin{bmatrix}I_L & 0\\ 0 & 1\end{bmatrix} \text{ and } \tau_3=\begin{bmatrix}1 & 0 & 0 & 0\\ 1 & 0 & 0 & 0\\ 1 & 0 & 0 & 0\\ 0 & 0 & 0 & -1\end{bmatrix}=\begin{bmatrix}I_L & 0\\ 0 & -1\end{bmatrix},$$

where I_L is a [3 × 3] minor matrix, introduced in (Darvas 2015). (E is the unit element of the HySy group.)

According to the above-mentioned derivations, in the presence of a velocity-dependent field, one should apply the extended R transformation matrix (cf., (10.4), (10.7)) for the matrix of the Lorentz transformation. Let us introduce a [3 × 3] minor matrix in R, to get R in the form

$$R = \begin{bmatrix} R^{(M3)} & 0 \\ 0 & 1 \end{bmatrix}$$

where (cf., (10.6)):

$$R^{(M3)} = \begin{bmatrix} \cos\vartheta + (1-\cos\vartheta)u_1^2 & (1-\cos\vartheta)u_1u_2 - u_3\sin\vartheta & (1-\cos\vartheta)u_1u_3 + u_2\sin\vartheta \\ (1-\cos\vartheta)u_2u_1 + u_3\sin\vartheta & \cos\vartheta + (1-\cos\vartheta)u_2^2 & (1-\cos\vartheta)u_2u_3 - u_1\sin\vartheta \\ (1-\cos\vartheta)u_3u_1 - u_2\sin\vartheta & (1-\cos\vartheta)u_3u_2 + u_1\sin\vartheta & \cos\vartheta + (1-\cos\vartheta)u_3^2 \end{bmatrix} \tag{10.9}$$

Transformations $T^{(D)}$ in a velocity-dependent field (**D**) under the combination of the extended Lorentz transformation and the transformation matrices of the HySy take the following forms:

$$T_+^{(D)} = E \cdot R = \begin{bmatrix} I_L & 0 \\ 0 & 1 \end{bmatrix} \begin{bmatrix} R^{(M3)} & 0 \\ 0 & 1 \end{bmatrix} = \begin{bmatrix} I_L \cdot R^{(M3)} & 0 \\ 0 & 1 \end{bmatrix}$$

and (10.10)

$$T_-^{(D)} = \tau_3 \cdot R = \begin{bmatrix} I_L & 0 \\ 0 & -1 \end{bmatrix} \begin{bmatrix} R^{(M3)} & 0 \\ 0 & 1 \end{bmatrix} = \begin{bmatrix} I_L \cdot R^{(M3)} & 0 \\ 0 & -1 \end{bmatrix}$$

Now, we can formulate the sought-after transformation of the field (convolution of a traditional SM field and the associated non-SM **D** field) that is expected to eliminate 'unwanted' masses of the quanta (δ) of the **D** field. Note that in contrast to the fix mass of all SM bosons, the mass of δ depends on the relative velocity between the two interacting isotopes of the concerned field-charges. Therefore, we are expecting a transformation formula depending on velocity (Darvas, 2020b).

There are two bosons mediating interactions in these fields. There appears one of the SM (plus gravity) bosons, depending on the kind of interaction that we denote by a general character ξ. (ξ may denote either the graviton or the photon, one of the weak vector bosons, or one of the strong gluons.) There appear also the bosons of the **D** field, δ.

Thus, the transformation of their fields may take the following two forms in each of the respective interactions:

$$\begin{bmatrix} \mathbf{D}' \\ \mathbf{X}_{SM}' \end{bmatrix} = T_+^{(D)} \begin{bmatrix} \mathbf{D} \\ \mathbf{X}_{SM} \end{bmatrix} = \begin{bmatrix} I_L \cdot R^{(M3)} & 0 \\ 0 & 1 \end{bmatrix} \begin{bmatrix} \mathbf{D} \\ \mathbf{X}_{SM} \end{bmatrix} \tag{10.11}$$

and

$$\begin{bmatrix} \mathbf{D}' \\ \mathbf{X}_{SM}' \end{bmatrix} = T_-^{(D)} \begin{bmatrix} \mathbf{D} \\ \mathbf{X}_{SM} \end{bmatrix} = \begin{bmatrix} I_L \cdot R^{(M3)} & 0 \\ 0 & -1 \end{bmatrix} \begin{bmatrix} \mathbf{D} \\ \mathbf{X}_{SM} \end{bmatrix} \tag{10.12}$$

We can see that these transformations do not affect any of the SM bosons ($\mathbf{X'}_{SM}$ coincide with $\mathbf{X}_{SM}$, ξ' are equal to ξ). The latter are not subject to any transformation in the **D** field.[10.7] It is reassuring that the IFC model does not destroy the SM; it only extends that at very high velocities of the interacting agents. Eqs. (10.11) and (10.12) rotate the non-SM field **D** of the massive intermediate bosons and one of the SM fields $\mathbf{X}_{SM}$ to produce Goldstone bosons consisting of massless IFC bosons in **D'**, and the respective SM bosons in $\mathbf{X'}_{SM}$(cf., footnote 10.4).

The transformation of the **D** section of the field is the same both in (10.11) and (10.12)

$$\mathbf{D}' = I_L \cdot R^{(M3)} \cdot \mathbf{D}.$$

Since $I_L \cdot R^{(M3)}$ is a [3 × 3] matrix, this can be written as

$$\begin{bmatrix} \mathbf{D}_1' \\ \mathbf{D}_2' \\ \mathbf{D}_3' \end{bmatrix} = I_L \cdot R^{(M3)} \begin{bmatrix} \mathbf{D}_1 \\ \mathbf{D}_2 \\ \mathbf{D}_3 \end{bmatrix}$$

Let us write the transformation of this section of the field in detail. First, investigate the $I_L \cdot R^{(M3)}$ product:

$$I_L \cdot R^{(M3)} = = \begin{bmatrix} 1 & 0 & 0 \\ 1 & 0 & 0 \\ 1 & 0 & 0 \end{bmatrix}$$

$$\cdot \begin{bmatrix} \cos\vartheta + (1-\cos\vartheta)u_1^2 & (1-\cos\vartheta)u_1u_2 - u_3\sin\vartheta & (1-\cos\vartheta)u_1u_3 + u_2\sin\vartheta \\ (1-\cos\vartheta)u_2u_1 + u_3\sin\vartheta & \cos\vartheta + (1-\cos\vartheta)u_2^2 & (1-\cos\vartheta)u_2u_3 - u_1\sin\vartheta \\ (1-\cos\vartheta)u_3u_1 - u_2\sin\vartheta & (1-\cos\vartheta)u_3u_2 + u_1\sin\vartheta & \cos\vartheta + (1-\cos\vartheta)u_3^2 \end{bmatrix} =$$

$$= \begin{bmatrix} \cos\vartheta + (1-\cos\vartheta)u_1^2 & (1-\cos\vartheta)u_1u_2 - u_3\sin\vartheta & (1-\cos\vartheta)u_1u_3 + u_2\sin\vartheta \\ \cos\vartheta + (1-\cos\vartheta)u_1^2 & (1-\cos\vartheta)u_1u_2 - u_3\sin\vartheta & (1-\cos\vartheta)u_1u_3 + u_2\sin\vartheta \\ \cos\vartheta + (1-\cos\vartheta)u_1^2 & (1-\cos\vartheta)u_1u_2 - u_3\sin\vartheta & (1-\cos\vartheta)u_1u_3 + u_2\sin\vartheta \end{bmatrix} \tag{10.13}$$

10.7 The Weinberg angle mixes two bosons, both appearing in an SM interaction field. The CKM angles mix quark flavours in another, but also SM field. The HySy rotation angle does not mix the δ bosons of the **D** field with any of the SM bosons (denoted by ξ), what latter appear simultaneously in one of the SM interaction fields ($\mathbf{X}_{SM}$). Therefore HySy *does not mix* them, instead, it is expected to *rotate* the δ boson's field while leaving the respective ξ SM boson's field unchanged. The rotation formula to be derived in the following part of the paper corresponds to this expectation.

One sees that the three rows of the resulting matrix coincide ($\mathbf{D}_1' = \mathbf{D}_2' = \mathbf{D}_3' = \mathbf{D}'$) as expected. Thus:

$$\mathbf{D}' = \left[\cos\vartheta + (1-\cos\vartheta)u_1^2 + (1-\cos\vartheta)u_1u_2 - u_3\sin\vartheta + (1-\cos\vartheta)u_1u_3 + u_2\sin\vartheta\right]\mathbf{D}$$

$$\text{or}\quad \mathbf{D}' = \left[\cos\vartheta + (1-\cos\vartheta)u_1(u_1+u_2+u_3) + \sin\vartheta(u_2-u_3)\right]\mathbf{D} \qquad (10.14)$$

In order to discuss this value, let us introduce polar co-ordinates for the u_i unitary projected velocity components of the IFCS. Note that we still have prescribed no constraint for the axes in the configuration space, where the u_i velocity component projections point. We are free to orient these axes arbitrarily. Now, let us define the axes so that the inclination angle of $\boldsymbol{v}$ in respect of u_3 be Θ_D and, let ψ denote the rotation angle around u_3 in the u_1_u_2 plane that is perpendicular to u_3.

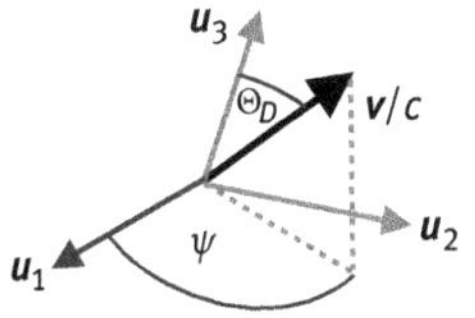

Figure 10.1

Since u_i are components of a unitary vector, the unitary projections of $\boldsymbol{v}$ are

$$u_1 = \sin\Theta_D\cos\psi$$
$$u_2 = \sin\Theta_D\sin\psi$$
$$u_3 = \cos\Theta_D$$

As it can be read from Fig. 10.1, Θ_D is the angle of a precession of $\boldsymbol{v}$ in a fixed reference frame. Θ_D characterises the velocity dependence of the transformation of the field $\mathbf{D}$. Now:

$$\mathbf{D}' = \begin{bmatrix} \cos\vartheta + (1-\cos\vartheta)(\sin^2\Theta_D\cos^2\psi + \\ +\sin^2\Theta_D\cos\psi\sin\psi + \sin\Theta_D\cos\psi\cos\Theta_D) + \\ +\sin\vartheta(\sin\Theta_D\sin\psi - \cos\Theta_D) \end{bmatrix}\mathbf{D}$$

One can fix the reference frame, considering that the precession of $\boldsymbol{v}$ around the axis u_3 cannot depend on the phase angle (of a rotation by ψ) in the u_1–u_2 plane. One can interpret ψ as a phase parameter of the spontaneous symmetry breaking in the $\mathbf{D}$ field. So, we are free to fix ψ (by an arbitrary choice) as $\psi = \pi/2$. In this case, $\cos\psi = 0$, $\sin\psi = 1$. With the above assumptions on the orientation of the reference frame of the velocity components:

$$\mathbf{D}' = [\cos\vartheta + \sin\vartheta(\sin\Theta_D - \cos\Theta_D)]\mathbf{D}$$

Considering the identity $\cos\Theta_D - \sin\Theta_D \equiv \sqrt{2}\cos(\Theta_D + (\pi/4))$:

$$\mathbf{D}' = \left[\cos\vartheta - \sin\vartheta\sqrt{2}\cos\left(\Theta_D + \frac{\pi}{4}\right)\right]\mathbf{D}$$

The transformation of **D** is a function of the symbolic angle ϑ and Θ_D. Both can be expressed with the relative velocity of the interacting field-charges. ϑ is defined by the Lorentz transformation, Θ_D by the transformation in the HySy field **D**. In simpler form:

$$\mathbf{D}' = \left[\sqrt{1 - \left(\frac{v}{c}\right)^2} - \sqrt{2}\frac{v}{c}\cos\left(\Theta_D + \frac{\pi}{4}\right)\right]\mathbf{D} \tag{10.15}$$

Inserting this in (10.11) [and in (10.12), respectively]:

$$\begin{bmatrix} \mathbf{D}' \\ \mathbf{X}_{SM}' \end{bmatrix} = \begin{bmatrix} \sqrt{1 - \left(\frac{v}{c}\right)^2} - \sqrt{2}\frac{v}{c}\cos\left(\Theta_D + \frac{\pi}{4}\right) & 0 \\ 0 & \pm 1 \end{bmatrix} \begin{bmatrix} \mathbf{D} \\ \mathbf{X}_{SM} \end{bmatrix} \tag{10.16}$$

According to (10.16), ϑ and Θ_D define together the transformation that eliminates unwanted masses produced by the spontaneous symmetry breakdown in the **D** field, and justify the mass of the δ boson. This formula complies with the transformation of the electroweak field by the Weinberg mixing angle. However, there are differences as well.

First, (10.16) transforms two coupled fields together, one of which is not a SM field.

Second, while there are fixed mass bosons in the weak interaction, the mass of the HySy field's boson depends on the relative velocity of the interacting field-charges. This is in keeping with the velocity dependence of the **D** field and is reflected in the field's transformation formula for the elimination of unwanted masses produced by the spontaneous symmetry breakdown.

Third, we must remark that the assumption of IFC originates in the asymmetry expressed in the Møller scattering matrix (cf., Section 3). That assumption involved the mass difference between the IFC siblings. A later formula obtained in the SM for the Møller scattering asymmetry for electrons includes the (weak mixing) Weinberg angle. (The weak mixing angle explains only the surplus of the Z^o boson. The BEH mechanism gives account on the full mass of Z^o.) The value of the Weinberg angle varies depending on the momentum transfer. The momenta affect the fixed masses of the related weak bosons. In contrast to that, although the transformation formula in the **D** field affects also the mass of the quanta of the field, but it leads to a boson mass depending on the relative velocity of the particles between which it mediates. Moreover, the appearance of the velocity-dependent angle in the formula for the transformation of the **D** field is simpler than in the Møller scattering asymmetry.

At the same time, the angle in (10.16) runs over a wider scale than the Weinberg angle does.

In short, Eq. (10.16) is the formula by which spontaneous symmetry breaking transforms the respective quanta of the original SM field and of the **D** field.

We can expect that the (10.11)–(10.12) rotation matrices in the **D**–$\mathbf{X}_{SM}$ field couple (as expressed in (10.16)) in a rotation matrix form

$$\begin{bmatrix} \sqrt{1-\left(\frac{v}{c}\right)^2} - \sqrt{2}\frac{v}{c}\cos(\Theta_D + \frac{\pi}{4}) & 0 \\ 0 & \pm 1 \end{bmatrix} = \begin{bmatrix} \cos\varphi(v,\Theta_D) & \sin\varphi(v,\Theta_D) \\ -\sin\varphi(v,\Theta_D) & \cos\varphi(v,\Theta_D) \end{bmatrix} \tag{10.17}$$

where $\varphi(v,\Theta_D)$denotes an angle that mixes the **D** and the respective $\mathbf{X}_{SM}$ fields. However, the inclination angle Θ_D appears to be more characteristic for the rotation of the **D** field.

10.3.3 Spontaneous breaking point of HySy

According to (10.17) $\sin\varphi(v,\Theta_D)= 0$ involves meaning stable values for $\varphi(v,\Theta_D)$, while v and Θ_D may vary. Since $\sin\varphi(v,\Theta_D)= 0$,

(a) $\varphi = 0$ and $\cos\varphi = 1$; or

(b) $\varphi = \pi$ and $\cos\varphi = -1$.

In case (a): $\varphi = 0$, there occurs no transformation in the field **D**. In this case (cf., (10.11)), $T_+^{(D)}$ turns into the identity transformation: $T_+^{(D)} = \begin{bmatrix} 1 & 0 \\ 0 & 1 \end{bmatrix}$.

The case (b): $\varphi = \pi$ corresponds to a real transformation, by the matrix τ_3. According to (10.12) and (10.17)

$$\cos\left(\Theta_D + \frac{\pi}{4}\right) = \frac{\sqrt{1-(v/c)^2}+1}{\sqrt{2}(v/c)} \tag{10.18}$$

The formula in (10.18) provides limits for the domain of interpretation of v/c, and respectively, for Θ_D. One must avoid having the right side becoming larger than the value allowed by a cosine function. Discussion of these limits allows us to define the limits where HySy is broken. At first, we exclude Θ_D outside the domain $-(\pi/2) \le \theta_D \le (\pi/2)$ (otherwise the projection of v would point in the opposite direction than v_3), and we also consider that v/c runs from 0 to 1.

The negative value of the square root in the numerator in (10.18) provides either – excluded – precession angles less than $-\pi/2$ for Θ_D, or positive precession angles. Furthermore, for $\varphi = \pi$ rotation of the **D**–$\mathbf{X}_{SM}$ fields causes a sign inversion in the $\mathbf{v}$–Θ_D plane [flips the (velocity) vectors in these fields over in the opposite direction], only negative Θ_D precession angles can be interpreted: $-(\pi/2) \leq \theta_D \leq 0$. Thus, all precession angles provided by negative square roots in the numerator should be excluded.

Considering positive values of the numerator, expression (10.18) is meaningless when $(v/c) < \sqrt{\frac{8}{9}} \cong 0.943$. This is the minimal velocity where HySy prevails. Below this limit velocity $v < \sqrt{\frac{8}{9}}c$, the HySy is broken. At the critical $v = \sqrt{\frac{8}{9}}c$, the precession angle $\Theta_D = -(\pi/4)$. Starting from this value of the spontaneous symmetry breaking angle value, while the velocity (kinetic energy) increases further, the value of the Θ_D precession angle spontaneously bifurcates. It either increases, reaching $\Theta_D = 0$ at $v = c$; or decreases, reaching $\Theta_D = -(\pi/2)$ at $v = c$. The observable domain of Θ_D varies between $-(\pi/2) < \theta_D < 0$ (cf., Figures (10.2) and (10.3)).

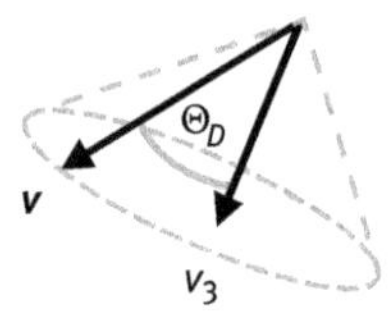

Figure 10.2

In other words, according to (10.17), the HySy rotation angle is $\varphi = \varphi(v, \Theta_D)$, as learned in the SM. The identity transformation (case (a)) indicates no transformation of the field **D** in SM terms. This case says, field **D** is present at the range $0 < v \leq c$, but its presence does not guarantee that HySy phenomena, like an HySy boson (dion), can be observed.

There may show up domains where HySy is broken. The real transformation (case (b)) indicates a π angle rotation of the plane of the fields $\mathbf{D}$-$\mathbf{X}_{SM}$, that is, vectors flip over in opposite direction. This justifies negative precession angles around velocity vectors in the **D** field.

The fixed value of the φ rotation angle in the SM hides the essence of the rotation of a field beyond the SM, like **D**. Since **D** exists beyond the SM, new rules may prevail for it. The essential characteristic angle is hidden in the velocity dependence of $\varphi(v, \Theta_D)$. The spontaneous symmetry breaking in HySy is characterised by that Θ_D precession angle. The curve of v/c in the function of the available values for Θ_D shows a sombrero-like graph (cf., Fig. 10.3). This complies with similar shapes for SM spontaneous symmetry breakings. According to the discussion of the set of values for the Eq. (10.18), Θ_D is interpreted between $-\pi/2$ and 0 (indicated as a bold line), and the respective values of velocity between $\sqrt{8/9} \leq \frac{v}{c} < 1$, at least, considering the positive value of the square root in the numerator in (10.18). The $\sqrt{8/9} \leq \frac{v}{c}$ limit for v/c means the limit under which velocity the HySy is broken.

In point of fact, HySy is effective between velocities $\sqrt{8/9} \leq \frac{v}{c} < 1$ with a domain of the precession angle $-(\pi/2) < \Theta_D < 0$. (We exclude the boundaries of the domain, because no massive particle can appear at velocity c.)

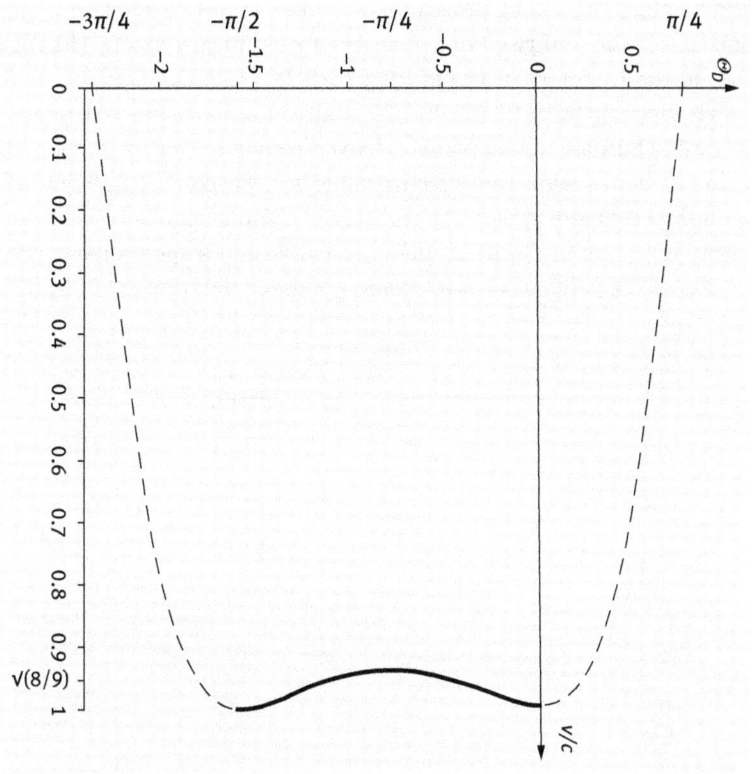

Figure 10.3: Real velocity.

The trigonometric formula demonstrates spectacularly the *precession* of the **D** field depending on the velocity (cf., Eq. (10.15)). The angle of the precession expresses the relation of $\boldsymbol{v}$ to $\boldsymbol{v}_3$. It can be shown clearly that the $\boldsymbol{v}$ vector of the velocity of one of the interacting IFC – with respect to the other field-charge – precesses around the straight, marked out by $\boldsymbol{v}_3$. The direction of $\boldsymbol{v}_3$ points between the two interacting field-charges and does not coincide with the direction of the real velocity $\boldsymbol{v}$ (cf., Fig. (10.1)). In another view, the $\boldsymbol{v}$ velocity vector precesses around the projection of the third component of the unitary length IFCS (Δ) from the **D** field to the configuration space. The angle of this precession changes with the change of the respective velocity $\boldsymbol{v}$. It can be also shown that the vector of velocity is always tangential to the line connecting the interacting agents, but this line is curved in the gauge field **D** induced by their own velocity. The latter can be expressed by the inverse of the matrix in Eq. (10.16). (Darvas, 2020a)

The vector $\boldsymbol{v}$ is interpreted in the velocity-dependent gauge field **D**. In the chosen orientation of the reference frame in **D**, the direction of its projection to the third axis, marked by $\boldsymbol{u}_3$ (cf., Fig. 10.1), coincides with the direction of the IFCS. Fig. 10.2

illustrates an angle-preserving projection of $\boldsymbol{v}$ and its third component $\boldsymbol{v}_3$ in the configuration space. The configuration space vector $\boldsymbol{v}$ (delineated in Fig. (10.3)), which is tangential to the trajectory of the moving particle, is its *real velocity*. At the same time, the vector component $\boldsymbol{v}_3$, inclined by angle Θ_D in respect to $\boldsymbol{v}$, defines the *effective velocity* of a boson-emitting particle in the direction towards a target particle that it enters in interaction with. This effective value is $v_3 = v \cdot \cos\Theta_D$ (cf., Figure 10.2). $\boldsymbol{v}_3$ is the longitudinal component of $\boldsymbol{v}$. The transversal component of $\boldsymbol{v}$ ($v \cdot \sin\Theta_D$) arrows in an arbitrary direction in the $\boldsymbol{u}_1$–$\boldsymbol{u}_2$ plane, according to a spontaneous precession. Fig. 10.4 shows the value of $\boldsymbol{v}_3$ as a function of the inclination angle Θ_D.

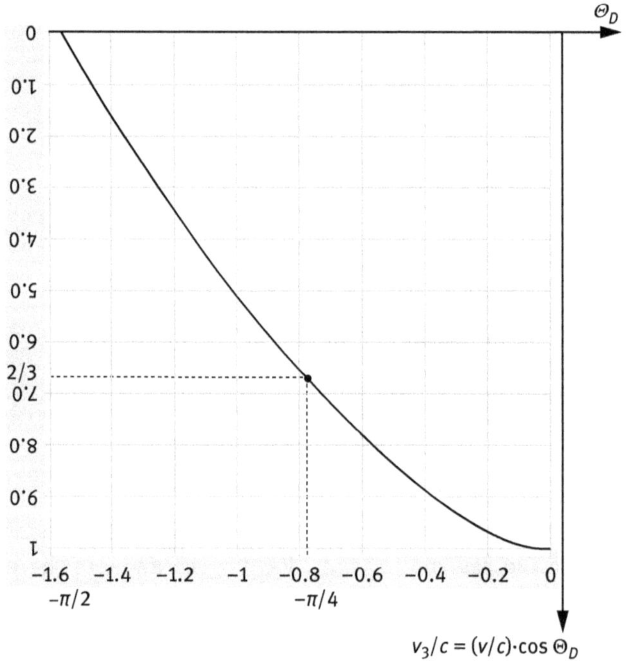

Figure 10.4: Effective velocity.

According to Fig. (10.3) (as a result of Eq. (10.18)), the HySy enters the scene at the velocity appearing by the angle $\Theta_D = -\pi/4$. As it can be read from Fig. (10.4), the effective velocity $\boldsymbol{v}_3$ becomes $(2/3)c$ at that point. This means that one can observe HySy above the *effective velocity* $\boldsymbol{v}_3 = (2/3)c$.

10.3.4 The mass of the mediating boson δ in light of the transformation of the D field

The set of the Θ_D angle values obtained determines the rotation of the **D** field that eliminates unwanted masses produced by the spontaneous symmetry breaking that is responsible for the mass of the field's δ boson. There is easy to see that the velocity dependence of Θ_D (Eq. (10.18)) is in close relation with the velocity-dependent coefficient (κ) of the mass of the δ boson.

As we saw, $m(\delta) = m_T - m_V = (\kappa - 1)m_V$ where the value of m_V is equal to the rest mass. At the minimum energy of the appearance of the HySy $\frac{m(\delta)}{m_V} = \kappa - 1 = \frac{1-\sqrt{1-(v/c)^2}}{\sqrt{1-(v/c)^2}} = 2.$ This energy value corresponds to the middle apex in the sombrero curve at $-\pi/4$ (cf., Fig. 10.3). Accordingly, the m_T mass of the Lorentz boosted IFC at the lower limit of HySy should be $(m_T/m_V) = \kappa = 3$. In other words, HySy is broken until the mass of the respective mediating boson does not reach the double of the rest mass of the emitting particle, or, what is the same, the Lorentz boosted mass does not reach the triple of the rest mass of the emitting particle. This expresses the lower limit of the *observability* of a boson δ that appears at velocity $v = \sqrt{\frac{8}{9}}c$.

When $\boldsymbol{v}$ approaches to zero, the mass of δ approaches to 0. However, δ cannot be observed near to such a low velocity, due to the spontaneous symmetry breaking of the IFCS field. This is expected, since low velocity means to return to the full domination of the SM. A measurable value is not allowed to appear for the **D** field's strength, and transformation of a boson in the **D** field within the limits of the SM. This means, the calculation confirms that one can observe no δ bosons when a **D** field vanishes. (Moreover, we showed in the previous paragraphs that there exists a stronger exact limit for v in order to eliminate the observation of a massive δ.) This fact confirms that the IFC theory extends the SM so that the SM is left intact and holds at the range of its validity, that is, at not extremely high energies.

We were seeking to find a transformation of the **D** field that may eliminate the mass of the δ boson.[10.8] This is equivalent to a rotation of the field demanded by the spontaneous symmetry breaking and a precession of the velocity $\boldsymbol{v}$ around its third projection in the **D** field (that produced the mass of the field's bosons).

10.3.5 Conclusions on the dion mass and the transformation of the D field

We have derived the transformation formula that eliminates 'unwanted' masses produced by the spontaneous symmetry breaking in the **D** field and justified the

10.8 Let us avoid confusing the identity transformation of the boson δ at zero velocity with the field rotation transforming its mass into 0.

mass of the quantum of the **D** field. The derivation justifies that **D** must be a gauge field, that is, velocity dependence cannot be considered a simple rotation in the configuration space defined in the matter field. Sections 5.2.5 and 5.2.8 showed that this **D** field is subject to an invariance under rotations of an *SU*(2) isomorphic group that characterises HySy. We demonstrated that the (IFCS) transformation in the **D** field must be coupled with a SM interaction field, and also that the transformation leaves the mediating bosons of the respective SM field intact. The derived formula confirms that the **D** field causes no observable effects at low velocities, but it should be taken into account at relativistic high velocities: it extends the SM but does not influence it in the range of its validity.

We derived a limit velocity $v/c = 2\sqrt{2}/3$, below which HySy is definitely broken. The Lorentz invariance is extended over this limit velocity (energy) by an invariance under HySy. A non-SM transformation of the **D** field interpreted by the BEH mechanism and discussed in Section 10.3.2 justifies the mass of the quanta of the field. This transformation is characterised by a mixing angle π and a precession inclination angle $\Theta_D(v)$. This inclination angle of the precession of vectors, interpreted in the velocity-dependent field, rotates the field that is responsible for eliminating the mass of the field's intermediate boson. The latter angle is interpreted by HySy, beyond the SM. The mass of the quanta of the **D** field (δ) depends on $\Theta_D(v)$.

11 IFC INTERACTIONS IN SM FIELDS

11.1 Mechanism of commuting *Δ*

An interaction between field-charges ℸ can be described in the standard model (SM) by a Feynman diagram as demonstrated in Figure 11.1:

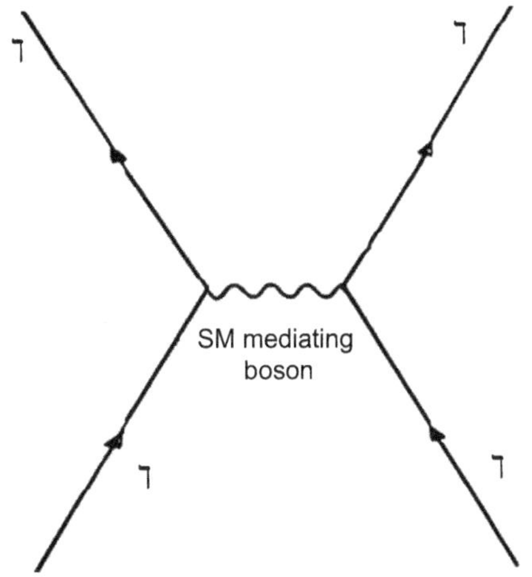

Figure 11.1: Single boson exchange in interaction between field-charges.

Here ℸ can mark the field charge of any physical interaction field. In the presence of a kinetic field **D**, the interacting particles have two properties present simultaneously: ℸ and *Δ*, and they exchange two intermediate bosons simultaneously: the one, characteristic to the given physical SM interaction (photon, weak current bosons, gluons, and at any probability graviton), and according to the flip of the isotopic field-charge spin (IFCS), they exchange simultaneously δ as well. Feynman (1949, appendix D, figure 8a and 8b) predicted the possibility of the contribution of two virtual quanta to the interaction that can be applied in our case.

We have to mention that Goldstone, Salam and Weinberg (1962, p. 970) found also an option for two mediating bosons. They concluded – among other options – that "if part of the loss of symmetry is due to the choice of a nonivariant boson mass . . ., then there must appear a two-boson pole at zero mass in the propagator of Φ^2". In their conclusions, Goldstone anticipated the possibility that symmetry breaking may involve "an inextricable combination of gauge and space-time transformations". It is something similar (but not the same how they later attempted) what we were trying to develop in this work (cf., Sections 10.2–10.3 on the boson mass). Then they noticed that Weinberg "has developed a method of rewriting any Lagrangian in order to introduce fields for bound as well as 'elementary' particles." This latter distinction impressed encouraging the author to develop the present project.

https://doi.org/10.1515/9783110713183-011

Interaction with the exchange of intermediate bosons was characterised in the SM like two children in two boats throw a ball to each other with a high frequency: as soon as they receive the ball they pass it back to the other (see Fig. 11.2).

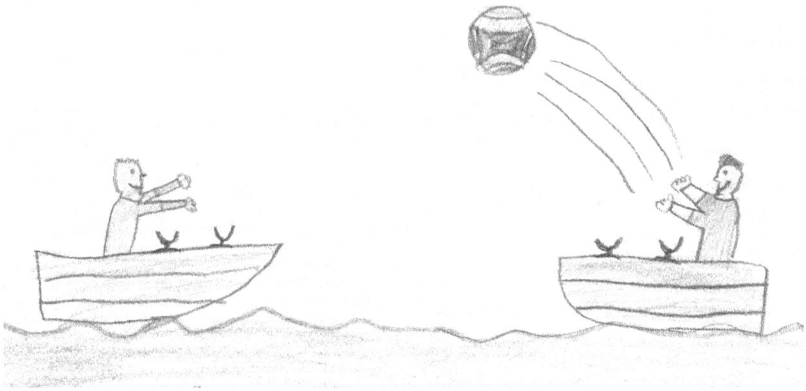

Figure 11.2: Single "boson" exchange. (Kristóf Sarkady, 8 y., 2010)

In the model proposed in the hypersymmetric (HySy) theory, the two children play the same game with two balls (see Fig. 11.3):

Figure 11.3: Two "boson" exchange. (Kristóf Sarkady, 8 y., 2010)

The interaction could be described with a Feynman diagram of the following type:

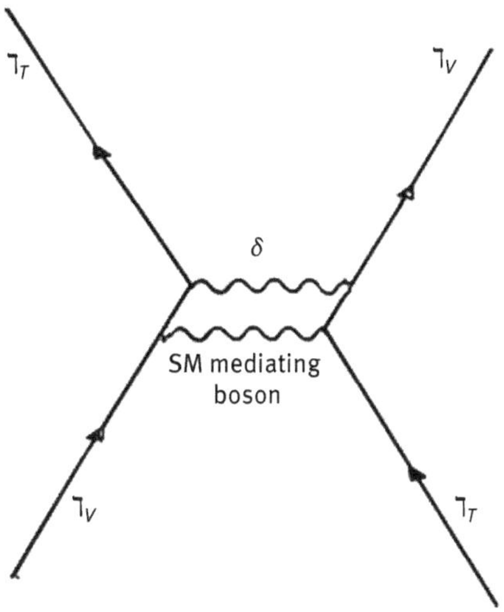

Figure 11.4: Two boson exchange in interaction between field-charges.

This mechanism introduces a loop in the chart which was unusual in the SM. (Cf., the 2008 preliminary results of experiments at the CERN on virtual-particle quantum loop effects, e.g. The Bell collaboration, 2008; Gershon, 2008; Wilczek, 2008; CERN workshop, 2008a, section 5; CERN workshop, 2008b; CERN workshop, 2008c, sections 4–5.) Nevertheless, making a distinction between potential and kinetic states, which correspond to observed facts, as well as the introduction of the corresponding kinetic gauge field were unusual too. Experiments must decide on the validity of the proposed model. *The criterion can be the detection of bosons to be identified as dions* (δ). Maybe virtual bosons assumed to mediate in box- or penguin-loop effects or other virtual bosons to be observed later could be identified with dions. Such identification depends on a prerequisite to replace the SUSY with the HySy model in the supporting theory.

All the rest in the IFCS (Δ) model is in accordance with experienced facts. They are only reinterpreted, which we called *another version of* the *facts*. This *version* predicted the existence of an invariance (of the property Δ) interpreted in a kinetic gauge field (**D**) whose quanta (δ) are assumed to mediate between the potential (ψ_V) and kinetic (ψ_T) states of the respective field-charges (ꓶ). We refer to this mechanism as the exchange of IFCS (or Δ) and the conservation of this property governed by HySy.

In the following sections, we will discuss the IFC interaction in the four fundamental physical interaction fields.

11.2 Hypersymmetry applied to gravitational interaction

First, we will discuss in this section, how the distinction between *gravitational* and *inertial masses* modifies physical equations. We saw in Section 2.4.1 that gravitational and inertial masses are – at least near to rest – *quantitatively* equivalent. At the same time, they are *qualitatively* different physical properties. We showed also that this difference has not been reflected in the traditional physical equations. It makes itself apparent at high velocities relative to the observer, where the two kinds of masses differ also in their quantities. These two kinds of masses were considered as *isotopic field-charges* (IFC) of the gravitational interaction field. The two, isotopic mass siblings are considered the field sources of the energy elements in the stress–energy tensor of the gravitational field and the kinetic (momentum-like) tensor elements, respectively. Taking into account this issue, this distinction between the mass siblings involved changes also in our general view on the physical world.

Putting isotopic masses in equations destroys their Lorentz invariance, and the symmetry of the $T_{\mu\nu}$ tensor. This would contradict our traditional precondition demanded from the physical theories, namely, a long-lasting paradigm that equations of the physical interactions must be subject to invariance under the Lorentz transformation (alone); moreover, it is not only a necessary but also a sufficient condition. The latter is not the case: no physical principle stated that Lorentz invariance is a sufficient condition demanded from the physical equations. There exist other invariances that may appear together and combined with Lorentz's. Yet, we would like to keep the symmetry of the stress–energy tensor.

Second, to *restore the* apparently lost *invariance*, we presented an additional invariance (a symmetry isomorphic with the *SU*(2) group) in Section 5.2. Later (Section 8.2), it was shown that IFC are subject to HySy (conservation of a property called the IFCS) that can transform them into each other (i.e. to rotate the IFC in an abstract IFCS field, where they can occupy two positions). It has been shown in Section 5.2 that HySy can be characterised by the help of an algebra (called *τ-algebra*). That symmetry guarantees to keep the covariance of our physical field equations and restores – among others – the violated symmetry of the stress–energy tensor. HySy is broken at lower velocities (lower kinetic energies). Therefore, at least near to rest, one can observe the two IFC of the gravitational field equivalent. This indistinguishability is formulated as an equivalence principle. However, as we saw in Section 2.3, equivalence – observed among limited conditions – does not mean identity. We put non-identical masses in physical equations, marking the gravitational and inertial masses by different notations. Since the gravitational mass appears in the potential (scalar) part of a Hamiltonian (*V*), we denoted it by $m_{\text{gravity}}=m_V$, while the inertial mass is the source of the kinetic (vector) part of a Hamiltonian (*T*), we denoted it by $m_{\text{inertia}}=m_T$. So, we modified the equations (including the gravitational) that led to novel conclusions at high velocities, while it did not result in changes near to rest. Our investigations include how the two isotopic masses behave at high kinetic energies.

Third, this physical description (in contrast to string theories) involves a new approach (among other interactions) to gravity. *Modification of the gravitational equation*[11.1] by inserting the two kinds of mass siblings in the stress–energy tensor results not only in destroying its Lorentz invariance but also in a few modifications in its solutions. Without going deep into mathematical details, this section directs shortly the attention to a few consequences and shows how the symmetry and invariance of the stress–energy tensor can be restored by HySy.

We will present here an *algebraic* way that restores the invariance of the modified stress–energy tensor under an extended gauge transformation (of the IFCS) by the tau algebra and the group composed of its elements, as introduced in Section 5.2.5.

The distinction between the two kinds of masses, $m_{\text{gravity}}=m_V$ and $m_{\text{inertia}}=m_T$ is not new in the history of physics. Their difference was known since Newton. However, due to their quantitative equivalence (at least at rest), it was not necessary to distinguish them qualitatively in the equations of physics. The experimental proof of their equivalence (first by R. Eötvös et al. 1910, 1922) and the formulation of the equivalence principle (by A. Einstein) inspired many textbook writers to identify them. This was pragmatically correct, but theoretically not justified. The problem became acute when technological development made possible to observe experimentally high energy interactions.

A few various approaches to handle the problem of the two kinds of masses in the last hundred plus years were discussed in Section 2.5. We indicated how the isotopic masses spoil the symmetry of the stress–energy tensor (cf., Figure 2.1). We premised that the apparently lost symmetry can be restored at high velocities by the introduction of an additional symmetry. We saw in Section 5.2.5 that the group of HySy and its representation by the τ algebra can fulfil that function.

11.2.1 Application of the τ algebra for the gravitational stress–energy tensor

Let us return to the distortion of the stress–energy tensor $T_{\mu\nu}$ (Figure 2.1) by the introduction of the isotopic masses m_V and m_T. Let $T_{\mu\nu}= [t''_{\mu\nu}]$ (see Eq. (11.1) below) and agree to tuck all constants into $t''_{\mu\nu}$, but not all parameters. We will put in those parameters and variables that are important in our aspect one by one. Observed from a moving reference frame, the $T_{\mu\nu}$ tensor components Lorentz transform by κ; let us tuck these κ also into $t''_{\mu\nu}$. The components T_{i4} can be expressed as $t''_{i4} ic\delta^{il} p_l(m_{Tl})$; and the components T_{4k} can be expressed as $t''_{4k}E_k(m_V)/c$. Furthermore, since the individual components of the stress–energy tensor denote densities, in the following we will

11.1 The *Int. J. Mod. Phys. D* published a series of papers between November 2012 and January 2014 on Modified Gravity Theories. Most of them concentrates on the left side of the Einstein equations ($F(R)$ gravity theories). A good review is given in Myrzakulov et al. (2013).

apply mass densities (ρ) instead of masses, and tuck the denominators also in the $t''_{\mu\nu}$-s. For simplicity, we hide the mass density dependence also into the t''_{ik}s in the stress density sector (i, k = 1, 2, 3). We are interested, first of all, in the components in the momentum density (T_{i4}) and the energy flux density (T_{4k}) sectors, therefore we write these components in more detail. Their individual components can be written as $t''_{i4}ic\delta^{il}p_l(\rho_{Tl})$ and $t''_{4k}E_k(\rho_V)/c$, respectively. Note that the respective icp_i and E_k/c tensor components denote qualitatively different quantities, not only due to the difference of the included m_V and m_T masses but also due to their transformation rules (cf., Hraskó, 2001, section 2.8, especially his eq. (2.8.3)). Here $p_i(\rho_{Ti})$ denotes that the p_i momentum density components depend on the respective inertial mass components. The index i in ρ_{Ti} denotes the individual mass component projections in the directions of the velocity of the moving reference frame. We must distinguish them because the same inertial mass m_T transforms by κ^3 in the direction of the velocity (longitudinal mass), and by κ in perpendicular directions (transversal mass). For we will not treat these different transformations in the following, we will also tuck these κ^3 and κ into $t'_{\mu\nu}$s. We will denote them in the following: $t''_{i4}icp_i(\rho_{Ti}) = t'_{i4}icp_i(\rho_T)$, where ρ_T denotes the rest value of the inertial mass density and is the same in all directions. We do not need to make similar tucking in the energy density flux since the gravitational mass densities (ρ_V) do not transform with the velocity, thus we can apply the following replacement, using $E_k/c=\rho_V c$: $t''_{4k}E_k(\rho_V)/c=t'_{4k}\rho_V c$. In order to distinguish them from the mass densities, we will denote the total energy density in t_{44} by index E (i.e. ρ_E).

Applying all these notations, the stress–energy tensor (2.1) in Section 2.6 can be now rewritten with the listed simplified notations in the following forms:

$$T_{\mu\nu} = \begin{bmatrix} t''_{11} & t''_{12} & t''_{13} & t''_{14}icp_1(\rho_{T1}) \\ t''_{21} & t''_{22} & t''_{23} & t''_{24}icp_2(\rho_{T2}) \\ t''_{31} & t''_{32} & t''_{33} & t''_{34}icp_3(\rho_{T3}) \\ t''_{41}E_1(\rho_V)/c & t''_{42}E_2(\rho_V)/c & t''_{43}E_3(\rho_V)/c & t''_{44}\rho_E(\rho_V) \end{bmatrix}$$

$$T_{\mu\nu} = \begin{bmatrix} t'_{11} & t'_{12} & t'_{13} & t'_{14}icp_1(\rho_T) \\ t'_{21} & t'_{22} & t'_{23} & t'_{24}icp_2(\rho_T) \\ t'_{31} & t'_{32} & t'_{33} & t'_{34}icp_3(\rho_T) \\ t'_{41}\rho_V c & t'_{42}\rho_V c & t'_{43}\rho_V c & t'_{44}\rho_E \end{bmatrix} \tag{11.1}$$

As mentioned, we would like to keep the symmetry of the stress–energy tensor. The expressions of the components in the momentum density and the energy flux density can be equal if the isotopic mass terms (or their densities) appearing in them can be transformed into each other. We must demonstrate that when transposing the tensor, the qualitatively different T_{i4} tensor components can be transformed into T_{4k} so that we get equal quantities. In other words, our goal is to show that there is a transformation which can transform the asymmetric T_{i4} and T_{4k} elements into each other.

Matrices of the group τ can provide this transformation. However, they do not transform directly the components of the stress–energy tensor, rather the eigenfunctions that satisfy quantum gravitational field equations (see Section 11.2.2). We see that the components in question differ in the mass density terms and their velocity dependence. To show that they can be transformed into each other, first, we must demonstrate that mass terms can be separated from other coefficients in the respective tensor elements.

To understand the relation, we illustrate this on a semi-classical example of virtual force components, expressed by the multiplication of the stress–energy tensor and a mass–density current $F^{\mu} = T_{\mu\nu}J^{\nu}$. Note that this force expresses the result of an interaction between a gravitational field with another mass' density current. The masses in the tensor and the current belong to two interacting systems; therefore, we will distinguish their mass densities with parenthetic upper indices (1) and (2). u^k denote contravariant velocity vector components of the mass–density current moving in respect to the gravitational field characterised by the tensor $T_{\mu\nu}$:

$$F^{\mu}_{virt} = T_{\mu\nu}J^{\nu} = \begin{bmatrix} t'_{11} & t'_{12} & t'_{13} & t'_{14}icp_1(\rho^{(1)}_T) \\ t'_{21} & t'_{22} & t'_{23} & t'_{24}icp_2(\rho^{(1)}_T) \\ t'_{31} & t'_{32} & t'_{33} & t'_{34}icp_3(\rho^{(1)}_T) \\ t'_{41}\rho^{(1)}_V c & t'_{42}\rho^{(1)}_V c & t'_{43}\rho^{(1)}_V c & t'_{44}\rho_E \end{bmatrix} \begin{bmatrix} \rho^{(2)}_{T1}u^1 \\ \rho^{(2)}_{T2}u^2 \\ \rho^{(2)}_{T3}u^3 \\ ic\rho^{(2)}_V \end{bmatrix} =$$

$$= \begin{bmatrix} t'_{11}\rho^{(2)}_{T1}u^1 + t'_{12}\rho^{(2)}_{T2}u^2 + t'_{13}\rho^{(2)}_{T3}u^3 - t'_{14}c^2p_1(\rho^{(1)}_T)\rho^{(2)}_V \\ t'_{21}\rho^{(2)}_{T1}u^1 + t'_{22}\rho^{(2)}_{T2}u^2 + t'_{23}\rho^{(2)}_{T3}u^3 - t'_{24}c^2p_2(\rho^{(1)}_T)\rho^{(2)}_V \\ t'_{31}\rho^{(2)}_{T1}u^1 + t'_{32}\rho^{(2)}_{T2}u^2 + t'_{33}\rho^{(2)}_{T3}u^3 - t'_{34}c^2p_3(\rho^{(1)}_T)\rho^{(2)}_V \\ t'_{41}c\rho^{(1)}_V\rho^{(2)}_{T1}u^1 + t'_{42}c\rho^{(1)}_V\rho^{(2)}_{T2}u^2 + t'_{43}c\rho^{(1)}_V\rho^{(2)}_{T3}u^3 + t'_{44}ic\rho_E\rho^{(2)}_V \end{bmatrix}$$

Now, let us introduce again a simplification: tuck the velocity dependence of the mass densities $\rho^{(2)}{}_{Ti}$ also in the $t'_{\mu k}$s ($k = 1, 2, 3$). They look with the shortened notation:$t'_{\mu k}\rho^{(2)}_{Ti} - t_{\mu k}\rho^{(2)}_{T}$. Now, we can disassemble this virtual force vector to the multiplication of a [4×4] matrix and a vierbein consisting of mass densities:

$$F^{\mu}_{virt} = \begin{bmatrix} t_{11}u^1 & t_{12}u^2 & t_{13}u^3 & t_{14}ic^2p^{(1)}_1 \\ t_{21}u^1 & t_{22}u^2 & t_{23}u^3 & t_{24}ic^2p^{(1)}_2 \\ t_{31}u^1 & t_{32}u^2 & t_{33}u^3 & t_{34}ic^2p^{(1)}_3 \\ t_{41}c\rho^{(1)}_V u^1 & t_{42}c\rho^{(1)}_V u^2 & t_{43}c\rho^{(1)}_V u^3 & t_{44}c\rho_E \end{bmatrix} \begin{bmatrix} \rho^{(2)}_T \\ \rho^{(2)}_T \\ \rho^{(2)}_T \\ i\rho^{(2)}_V \end{bmatrix} \tag{11.2}$$

The separability of the mass densities is demonstrated (at least in the case of the mass densities of the current). We saw in Section 5.2.2 that τ_2 can transform such vierbeins into each other:

$$\tau_2 \begin{bmatrix} \rho_T^{(2)} \\ \rho_T^{(2)} \\ \rho_T^{(2)} \\ i\rho_V^{(2)} \end{bmatrix} = \begin{bmatrix} \rho_V^{(2)} \\ \rho_V^{(2)} \\ \rho_V^{(2)} \\ i\rho_T^{(2)} \end{bmatrix} \tag{11.3}$$

and back.

We have any reason to assume that the mass dependence of the components of $T_{\mu\nu}$ is linear, and does not include either lower, or higher powers of m than 1, namely $t'_{\mu\nu}f(m_\lambda)$, where $f(m_\lambda)$ is a function of the respective isotopic mass. Provided that in (11.1) all $t'_{\mu\nu}(\rho^{(1)}{}_\lambda)$ (λ=T, V) have the linear multiplication form $t'_{\mu\nu}(\rho^{(1)}{}_\lambda) = t^{\#}{}_{\mu\nu}\rho^{(1)}{}_\lambda$, those $\rho^{(1)}{}_\lambda$ can also be detached from $T_{\mu\nu}$, and a similar τ_2 transformation applies to them.

According to the IFC theory, inertial masses interact always with gravitational ones, and *vice versa*. This holds in our mentioned example for the momentum density and the energy flux density sectors. Let us now discuss the rest of the tensor. The right-hand side of (11.2) can be written in the form:

$$\begin{bmatrix} t_{11}u^1 & t_{12}u^2 & t_{13}u^3 & t_{14}ic^2p_1^{(1)} \\ t_{21}u^1 & t_{22}u^2 & t_{23}u^3 & t_{24}ic^2p_2^{(1)} \\ t_{31}u^1 & t_{32}u^2 & t_{33}u^3 & t_{34}ic^2p_3^{(1)} \\ t_{41}cp_V^{(1)}u^1 & t_{42}cp_V^{(1)}u^2 & t_{43}cp_V^{(1)}u^3 & t_{44}cp_E \end{bmatrix} \begin{bmatrix} \rho_T^{(2)} \\ \rho_T^{(2)} \\ \rho_T^{(2)} \\ i\rho_V^{(2)} \end{bmatrix} =$$

$$= \begin{bmatrix} [t_{ik}u^k] & t_{i4}ic^2p_i^{(1)} \\ t_{4k}cp_V^{(1)}u^k & t_{44}cp_E \end{bmatrix} \begin{bmatrix} \underline{\rho}_T^{(2)} \\ i\rho_V^{(2)} \end{bmatrix}$$

$$\begin{bmatrix} [t_{ik}u^k] & t_{i4}ic^2p_i^{(1)} \\ t_{4k}cp_V^{(1)}u^k & t_{44}cp_E \end{bmatrix} \begin{bmatrix} \underline{\rho}_T^{(2)} \\ i\rho_V^{(2)} \end{bmatrix} =$$

$$= \begin{bmatrix} [t_{ik}u^k] & 0 \\ 0 & t_{44}cp_E \end{bmatrix} \begin{bmatrix} \underline{\rho}_T^{(2)} \\ i\rho_V^{(2)} \end{bmatrix} + \begin{bmatrix} 0 & t_{i4}ic^2p_i^{(1)} \\ t_{4k}cp_V^{(1)}u^k & 0 \end{bmatrix} \begin{bmatrix} \underline{\rho}_T^{(2)} \\ i\rho_V^{(2)} \end{bmatrix} \tag{11.4}$$

The second component of the sum in the right-hand side of (11.4) is in concordance with the IFC theory. The first component in the right-hand side of (11.4) includes the stress density and the full energy density sectors. The former contains stress components that, in principle by their names, are inertial, and are here to be multiplied with the inertial part of the current's mass density vector. The full energy density that depends, in principle, on the gravitational mass is to be multiplied with the gravitational part of the current's mass–density vector. However, these multiplications do not mean real interaction. They give additions to the value of the calculated virtual force, but do not represent the interaction itself. Moreover, due to shear, the stress section contains entangled contributions by the scalar and vector parts of the field strengths. The full energy density in T_{44} includes the contributions of all masses constituting the field, and is an invariant, that is, does not transform under interaction.

11.2.2 Hypersymmetry of the gravitational equations

The original gravitational equation is invariant under the Lorentz transformation, however, this property does not ensure sufficiency for the covariance of the equation under strongly relativistic conditions. As many publications state, the theory needs certain modifications (cf., footnote 11.1). Among others, Wüthrich et al. (2012) give a good review on approaches to quantum gravity (first of all, from the aspect of the "lost" time in some approaches to it). Covariance under the Lorentz transformation should be extended. Such an extension is discussed in this section. Section 8.4 showed that Δ (IFCS) is a two-valued, spin-like property that appears in the presence of a kinetic gauge field at high energies. To recall: The two isotopic states are associated with the field-charges appearing in the scalar (potential, V) and the vector (kinetic, T) parts of a Hamiltonian, respectively. A qualitative distinction was made between the two (isotopic) kinds of field-charges. A given particle can occupy either the one or the other IFC state.

The assumption that the wave function of a given particle may be in a potential state with amplitude ψ_V or a kinetic state with amplitude ψ_T applies also to the gravitational interaction.[11.2] The wave function of a single physical particle is an entanglement of the probabilities being in the potential or the kinetic state at a given moment: $\psi = \begin{pmatrix} \psi_T \\ \psi_V \end{pmatrix}$, where ψ_T is a three-component column.

In quantum gravity, ψ can be interpreted, among others, for example, as a subject of a formally Schrödinger-like but in fact Wheeler–deWitt equation (WdW, with Hamiltonian constraint) that contains information about both the geometry and the matter content of the universe. It is a functional that includes space–time and configurations of other physical fields over it. It is (re)parametrised not only by the space and the time parameters separated but also by the Δ parameters.

As regards the latter ones, we denoted the eigenfunctions ψ_T and ψ_V that belong to the two eigenvalues of the operator $\boldsymbol{\tau}$, respectively. The operators $\boldsymbol{\tau}$ should fulfil

11.2 According to one of the interpretations of the IFCS theory, the energy of a single particle at a given moment is considered to be ideally concentrated either in the scalar or in the kinetic part of

eigenvalue equations $\boldsymbol{\tau}\varphi = k\varphi$, where φ are eigenfunctions of the operator $\boldsymbol{\tau}$, and k are numbers. Two IFCS (Δ) positions belong to the sources of the two states, as shown in the previous sections. During the transition from ψ_V to ψ_T and back, the source of the field, that is, the respective field-charge, needs to change its Δ state. The operator $\boldsymbol{\tau}$, whose matrix algebra was presented in Section 5.2, affects those Δ states.

First, recall that χ and ϑ are the eigenfunctions of the Δ IFCS's $\boldsymbol{\tau}$ operators. (As mentioned, the corresponding eigenfunctions were denoted also by $\varphi_+^{(\tau)}$ and $\varphi_-^{(\tau)}$). The eigenvalue equations are $\tau\varphi_+^{(\tau)} = \frac{1}{2}\varphi_+^{(\tau)}$ or $\tau\chi = \frac{1}{2}\chi$ and $\tau\varphi_-^{(\tau)} = -\frac{1}{2}\varphi_-^{(\tau)}$ or $\tau\vartheta = -\frac{1}{2}\vartheta$, respectively. The eigenvalues of Δ can take $\pm(1/2)$. The operator $\boldsymbol{\tau}$ refers to the transformation matrices of the HySy *group* discussed in Section 5 that rotate, for example, the IFCS in a velocity-dependent kinetic gauge field. Since the operators responsible for the rotation in the kinetic gauge field do not affect the space- and time-dependent (and other dependence) components in the field equations, and *vice versa*, the particle's full state function ψ can be separated according to the following sum: $\psi = \psi'\varphi_+^{(\tau)} + \psi''\varphi_-^{(\tau)}$, where ψ' and ψ" denote the state functions affected by the space–time-dependent operators. The ψ state functions in Eq. (11.5) are functions of the x^ν space–time co-ordinates, as well as of the $\Delta = \pm(1/2)$ parameters (according to the set of eigenvalues of the $\boldsymbol{\tau}$ operators). The differential operators act only on $\psi(x^\nu)$, and the $\boldsymbol{\tau}$ operators on the eigenfunctions of Δ. This is a useful property because the doubled solutions can be obtained by extending the space–time dependent state functions with IFCS (Δ)-dependent functions as multipliers.

Thus, the $\psi = \psi(x^\nu, \Delta)$ state function can be dissociated to the following products according to the two eigenfunctions of the $\boldsymbol{\tau}$ matrix operators:

$$\psi(x^\nu, \Delta) = \psi'(x^\nu)\varphi_+^{(\tau)}(\Delta) + \psi''(x^\nu)\varphi_-^{(\tau)}(\Delta) \tag{11.5}$$

Eq. (11.5) involves also that the ψ state functions in each of the original Einstein equations' solutions can be separated into two further space–time dependent functions, according to the additional bivariant opposite positions of the IFCS. (It is important to notice that the coincidence of the set of matrices in the τ- and the T-algebras ensures that the number of solutions is *doubled* thanks to the IFCS invariance, according to the two new degrees of freedom brought in by the two possible positions of the IFCS – compared with the *quadruple* solutions of the Dirac equation due to the new degrees of freedom by the $\boldsymbol{\rho}$ and $\boldsymbol{\sigma}$ operators (Dirac, 1928, 1929).) It can be exemplified so that, according to the two isotopic states of the field-charges,

---

its Hamiltonian (cf., Sections 10.1.1.2 and 10.2). The corresponding state functions belonging to these two idealised reduced Hamiltonians are denoted by ψ_V and ψ_T. They characterise the s.c. *potential* and *kinetic* states of the particle. The particle takes one of these two states at certain *probabilities*. According to this interpretation, the particle can change its state (or oscillate) between the fully potential and the fully kinetic states. The observable state function can be characterised by the mixture of these two probability states and denoted by ψ. The IFCS theory describes transition between the two states.

two separated solutions are assigned to each of the known solutions of the gravitational equations. The ψ functions satisfy $\hat{H}\psi(x^\nu, \Delta) = 0$ where $\hat{H}$ is a WdW Hamiltonian (and in whose form the time dependence of ψ disappears). According to the IFCS theory, the graviton exchange, which is a result of the symmetry of the space–time part of the state function, must be accompanied by the exchange of another boson, the dion, due to the symmetry of the φ^τ part of the state function(al) under the transformations of the $\boldsymbol{\tau}$ group, which is isomorphic with the $SU(2)$ group. In short, they are invariant under HySy transformations. The dion must carry the mass difference between the gravitational and the Lorentz transformed inertial masses (cf., Section 10.3.1). We mention also that while the operators $\boldsymbol{\tau}$ may rotate the masses (mass densities) among themselves, they leave the rest of the state functions intact; moreover, they guarantee to save the real values in the stress–energy tensor, and the negativity of the T_{44} component.

Recall that the transformation of the field-charges in the abstract IFCS field is independent of the given physical interaction (here namely the gravitational); the same transformation rotates the field-charges of all fundamental interactions between their scalar (potential) and vector (kinetic) states appearing in the respective parts of their Hamiltonian – in an interaction-independent IFCS (or Δ) field. The gravitational eigenvalue equation does not lose its covariant character by its extension with the IFCS section, rather, this extension guarantees its covariance in the presence of a velocity-dependent field and IFC.

11.2.3 The affine connection field

Note that the $g_{\mu\nu}$ metric tensor appearing in the gravitational equations, and consequently, the affine connection field and the curvature tensor formed from its derivatives depend on space–time and velocity co-ordinates. With the appearance of the dependence on the velocity vector, the curvature becomes dependent on its direction in each space–time point. The direction (additional parameter) attributed to each space–time point is defined by the orientation of the velocity of a test unit-mass in the given space–time point, $\mathbf{v}/|v|$. The metric is defined by the dependence of $g_{\mu\nu}$ on $\dot{x}_\rho$ (and x_ρ). The field is no more subject of Riemann geometry, more likely a Finsler geometry. (On the application of a Finsler geometry cf., Darvas, 2012c–e.) In general (classical) case, the affine connection field (curvature):

$$\Gamma_{\lambda\mu\nu} = \frac{1}{2}\left[\partial_\mu g_{\lambda\nu} + \partial_\nu g_{\lambda\mu} - \partial_\lambda g_{\mu\nu}\right] \quad (\Gamma^\lambda_{\mu\nu} = g^{\lambda\rho}\Gamma_{\rho\mu\nu})$$

and the Ricci tensor:

$$R_{\mu\nu} = \partial_\mu \Gamma^\lambda_{\nu\lambda} - \partial_\lambda \Gamma^\lambda_{\mu\nu} + \Gamma^\lambda_{\mu\sigma}\Gamma^\sigma_{\nu\lambda} - \Gamma^\lambda_{\sigma\lambda}\Gamma^\sigma_{\mu\nu}$$

In our case, the Ricci tensor is

$$R_{\mu\nu} = \partial_\mu \Gamma^{\lambda}_{\nu\lambda} - \partial_\lambda \Gamma^{\lambda}_{\mu\nu} + \Gamma^{\lambda}_{\mu\iota}\Gamma^{\iota}_{\nu\lambda} - \Gamma^{\lambda}_{\iota\lambda}\Gamma^{\iota}_{\mu\nu} =$$
$$= \partial_\mu g^{\lambda\kappa}\Gamma_{\kappa\nu\lambda} - \partial_\lambda g^{\lambda\kappa}\Gamma_{\kappa\mu\nu} + g^{\lambda\kappa}\Gamma_{\kappa\mu\iota} g^{\iota\kappa}\Gamma_{\kappa\nu\lambda} - g^{\lambda\kappa}\Gamma_{\kappa\iota\lambda} g^{\iota\kappa}\Gamma_{\kappa\mu\nu}$$

where

$$\Gamma_{\lambda\mu\nu} = \frac{1}{2}\left[\frac{\partial g_{\lambda\mu}}{\partial \dot{x}^{\rho}}\partial_\rho \dot{x}^{\nu} + \frac{\partial g_{\lambda\nu}}{\partial \dot{x}^{\sigma}}\partial_\sigma \dot{x}^{\mu} - \frac{\partial g_{\mu\nu}}{\partial \dot{x}^{\tau}}\partial_\tau \dot{x}^{\lambda}\right]$$

Comparing this $R_{\mu\nu}$ with the covariant forms of $F^{(1)}$ and $F^{(2)}$ defined in Section 7.3, the similitude is obvious:

$$F^{(1)\nu\mu}_{\alpha}(\dot{x}) = \frac{\partial D_{\nu,\alpha}(\dot{x})}{\partial \dot{x}_\mu} - \frac{\partial D_{\mu,\alpha}(\dot{x})}{\partial \dot{x}_\nu} - igC^{\gamma}_{\alpha\beta} D_{\dot{\nu},\beta}(\dot{x}) D_{\dot{\mu},\gamma}(\dot{x})$$

$$F^{(2)\mu\nu}_{\alpha}(x) = \frac{\partial D_{\dot{\rho},\alpha}\Lambda^{\rho}_{\mu}}{\partial x_\nu} - \frac{\partial D_{\dot{\sigma},\alpha}\Lambda^{\sigma}_{\nu}}{\partial x_\mu} - igC^{\gamma}_{\alpha\beta} D_{\dot{\rho},\beta}\Lambda^{\rho}_{\mu} D_{\dot{\sigma},\gamma}\Lambda^{\sigma}_{\nu}$$

This velocity-dependent $R_{\mu\nu}$ Ricci tensor and $F^{(1)}$ should be replaced in the Einstein equation:

$$T_{\mu\nu} = -\frac{1}{8\pi G_N}\left(R_{\mu\nu} - \frac{1}{2}Rg_{\mu\nu} + \Lambda g_{\mu\nu}\right) = F^{(1)}{}_{\mu\lambda}F^{(1)}{}_{\lambda\nu} + \frac{1}{4}\delta_{\mu\nu}F^{(1)}{}_{\lambda\sigma}F^{(1)}{}_{\lambda\sigma}$$

Due to velocity dependence, the Schwarzschild solution cannot be applied! It must be replaced by a Finslerian solution.

11.2.4 The mechanism of the Δ exchange in gravitational interaction

The model sketched in Section 11.1 can be applied to the gravitational interaction (see Fig. 11.5) with the reservation that the unit-mass, as the source of the gravitational field, has not been identified, and there is also questionable its fermionic character since the graviton, as its mediating boson has spin 2. The latter assumption maintains restriction for the applicable form of the Lagrangian but does not question the applicability of the described model. Nevertheless, gravitational mass can be naturally associated with the potential state of mass (m_V), and inertial mass can be associated in the same way with the kinetic state of mass (m_T).

This interpretation provided specific insight into the centuries-old problem of the distinction and equivalence of the two masses. Graviton is responsible for the interaction between two elementary mass units, in the course of which the two actors are in two different initial states, and δ_G (G denotes that it is the dion which conveys the gravitational interaction) is responsible to switch both of them in the opposite Δ state as a result of the interaction. The process with the exchange of the two bosons, continues permanently.

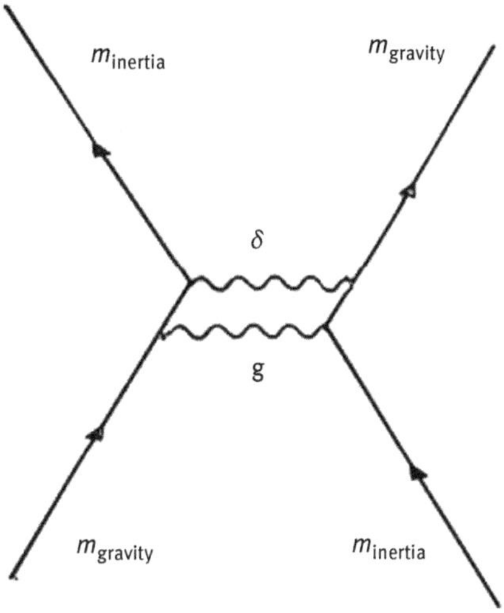

Figure 11.5: Two boson exchange in interaction between the field-charges of the gravitational field. (δ denotes the dion, g denotes the graviton.)

11.3 Hypersymmetry applied to electromagnetic interaction

Previous sections of this book presented a general field-theoretical model for the conservation of the IFCS, applicable to different kinds of interaction. It was based on the qualitative distinction between two types of the source charges (in particular, gravitational and inertial masses, Coulomb and Lorentz charges) of physical fields, and if so, it assumed interaction between the isotopes of the two types of field-charges.

The previous section presented the IFCS theory and its possible applications first to the description of gravitational interaction. The presence of a kinetic field, with a velocity-dependent metric and IFC (namely in that case: distinguished gravitational and inertial masses), required the application of a (velocity arrowed) direction-dependent, anisotropic (that means, Finsler) geometry (cf., Darvas, 2012a, e). Gravity as a Finslerian phenomenon has been discussed by other authors as well (e.g. Barletta and Dragomir, 2012; Duffy et al., 2012; see also 't Hooft, 2011, appendices A and B, although there without mentioning Finsler).

Before we apply the HySy algebra, we extend the original Dirac equation in analytical form in three consecutive steps to a further extension of the field-theoretic model of the electromagnetic interaction.[11.3] In Section 3, we introduced the isotopic electric

11.3 Electromagnetic field theories were related with anisotropic geometries in, at least, two terms. First, the Dirac matrices, introduced in QED (1928), follow the rules of hypercomplex numbers (as shown, among others, by Achiezer and Berestetskii (1969, p. 90)). Something similar has been introduced by the help of Clifford algebras in (Hiley and Callaghan 2012). Later Dirac published two

charges in the classical Maxwell EM theory. We showed there that applied alone, the presence of isotopic electric charges spoiled the Lorentz invariance of the Maxwell theory. To restore the violated invariance, we refer to the conservation of the IFCS, proven in Section 7.3, so that the two transformations, applied together, ensure the invariance and save the physical relevance of the theory. Now at first, we introduce the isotopic electric charges in the classical Dirac equation (Darvas, 2012c). Second, we extend the Dirac equation with a kinetic field, introduced in Section 6.2. The main part of the section discusses this extended equation. The discussion of the Dirac theory was treated also by de Haas (2004b); nevertheless, he did it without this kinetic extension.

If one observes certain formal similarity to the introduction of the A_μ and B_μ gauge fields (Weinberg, 1967), when we introduce a D_μ gauge field, as well as to the discussion of the proportionality of the g and g' coupling constants to the electron charge – that is not by chance. Nevertheless, the similitude is merely formal. We would like to underline here the differences. The D_μ gauge field in this theory differs from Weinberg's B_μ gauge field in its physical interpretation (cf., the derivation of the components of D_μ in Section 11.3.8.3.2). Concerning the proportionality between the two kinds of coupling constants/electric *charges*, in contrast to Weinberg (1967,) this book places the emphasis on the difference between the two kinds of *masses*. Otherwise, these minor alterations do not influence the conclusions (cf., e.g. 11.46 in Section 11.3.12).

11.3.1 Isotopic electric charges in classical EM

We saw that according to the introduced IFC theory, we can replace the electric charges appearing in our equations even in the classical EM by two different (isotopic) charges, a Coulomb-type one, and a kinetic-type one, although they destroyed the invariance of the equations under Lorentz transformation. The Coulomb-type charge is associated with the potential part of the Hamiltonian (V), and the kinetic type with the kinetic part of the Hamiltonian (T), and they appear in the components of a current density, respectively (cf., Section 3.5). Therefore, we demanded and then proved the existence of an extensional symmetry transformation. Now, as a next step, we introduce IFC in the derivation of the Dirac (1928) equation.

---

essential extensions to his QED theory (Dirac, 1929, 1951a, b, 1962). Second, Dirac (1962) introduced a curvilinear co-ordinate system. So, he defined an auxiliary co-ordinate system y_Λ, "which is kept fixed during the variation process and use functions $y_{\Lambda(x)}$ to describe the x co-ordinate system in terms of the y co-ordinate system". He defined the y system "so that the metric for the x system" be $g_{\mu\nu} = y_{\Lambda,\mu} y^{\Lambda}_{,\nu}$. This metric, which then appears in the Hamiltonian of the electromagnetic interaction, was the first step to a later Finsler extension. Some consequences have been discussed in Brinzei and Siparov (2008), Voicu (2010), Siparov (2012), and Pons et al. (2000). Concerning the parallel presence of a scalar and a kinetic gauge field, Sarkar et al. (2013) enlarge the configuration space by including a scalar field additionally, and by taking anisotropic models into account too.

11.3.2 Isotopic electric charges in QED

Dirac considered in first (unperturbed) approximation a case of no field when the wave equation reduces to

$$(-p_4^2 + p_i^2 + m^2c^2)\psi = 0 \tag{11.6}$$

where $p_4 = \mathbf{W}/c = ih(\partial/c\partial t)$ and $p_i = -ih(\partial/\partial x_i)$ (i=1, 2, 3) and the wave equation (11.6) should be written in the form $(\mathbf{H}-\mathbf{W})\psi=0$.

To maintain the required linearity of the Hamiltonian $\mathbf{H}$ in p_μ one introduces the dynamical variables α_i and β which are independent of p_μ, that is, they commute with x_i, t. Here Dirac considered particles moving in empty space so that all points in space were equivalent, and one can expect the Hamiltonian not to involve t and x_i. It follows that α_i and β are independent of x_i, t, that is, they commute with p_μ (μ=1, 2, 3, 4), although this latter held only until we did not distinguish gravitational and inertial masses. Dirac introduced his matrices intending to have other dynamical variables besides the co-ordinates and moments of the electron, so that α_i and β may be functions of them, and that the relativistic Lorentz invariant wave function depended on these variables. Dirac's wave equation took the form:

$$(p_4 + \alpha_1 p_1 + \alpha_2 p_2 + \alpha_3 p_3 + \beta)\psi = 0 \tag{11.7}$$

According to our assumptions, from here on we should modify the clue followed by Dirac. This equation must lead to a condition where we consider that the interacting two charges are carried by particles with masses in two IFC states, one of them in potential, the other in kinetic state. Since the masses of the carriers appear explicitly in β, we have to introduce two kinds of β, corresponding to the two isotopic mass states: β_V and β_T. We have to note that for the sake of relativistic invariance of the four-momentum's square, the mass square in Eq. (11.6) must be equal with the rest mass. We saw this is – at least numerically – equal with the potential (gravitational) mass: m_V=m_0. We make a qualitative distinction between the masses m_V and m_T, taking into account that the numerical value of the kinetic mass at relativistic velocities is

$$m_T = m_0 \Big/ \sqrt{1-(v^2/c^2)}$$

(where m_0 is the rest mass of the particle and v is the velocity of the particle in kinetic state relative to the interacting other particle in potential state). This qualitative distinction will obtain significance later. Thus, Eq. (11.7) leads to

$$(-p_4 + \alpha_1 p_1 + \alpha_2 p_2 + \alpha_3 p_3 + \beta_V)(p_4 + \alpha_1 p_1 + \alpha_2 p_2 + \alpha_3 p_3 + \beta_T)\psi = 0 \tag{11.8}$$

In order to agree with Eq. (11.6) in the form $(-p_4^2 + p_i^2 + m_V m_T c^2)\psi = 0$ – considering, in accord with Darvas (2011) that a particle in a kinetic state interacts always with another, which is in a potential state – we must demand that the coefficients fulfil the conditions:

(d1) $\alpha_i^2 = 1$

(d2) $\alpha_i\alpha_j + \alpha_j\alpha_i = 0 \quad i \neq j$

(d3) $\beta_V\beta_T = m_V m_T c^2$

(d4) $\alpha_i p_i \beta_T + \beta_V \alpha_i p_i = 0$

(d5) $\beta_V p_4 - p_4 \beta_T = 0$

(d1) and (d2) coincide with conditions established by Dirac. Conditions (d3)–(d5) do not follow from Dirac's original clue. We discuss them separately.

Considering commutations and introducing the notations $\beta_V = \alpha_4 m_V c$ and $\beta_T = \alpha_4 m_T c$ (and assuming that α_4 commutes with the scalar m_V) then from (d4) and (d3) follows:

(d3*) $\beta_V\beta_T = \alpha_4 m_V c \alpha_4 m_T c = \alpha_4^2 m_V m_T c^2 = m_V m_T c^2 \to \alpha_4^2 = 1$

and in accordance with Dirac, we get from (d1), (d3*), and (d2), (d4)

$$\alpha_\mu^2 = 1, \quad \alpha_\mu\alpha_\nu + \alpha_\nu\alpha_\mu = 0 \quad (\mu \neq \nu) \quad \mu, \ \nu = 1, \ 2, \ 3, \ 4$$

These coincide with the conditions deduced by Dirac (the only difference is the qualitative replacement of m^2 by $m_V m_T$, taking into account the earlier notice on the relativistic invariance of the four-momentum's square), and they involve that the derived Dirac matrices will not take different forms in our treatment either.

Condition (d5) deserves additional attention. Since there appear qualitatively different β_V and β_T we cannot commute β and p_4. We get

$$m_V \frac{\partial}{\partial t}\psi = \frac{\partial}{\partial t}(m_T \psi)$$

and from here

$$(m_T - m_V)\frac{\partial \psi}{\partial t} = -\left(\frac{\partial m_T}{\partial t}\right)\psi \tag{11.9}$$

If m_T is stationary, then either ψ is stationary too, or $m_V = m_T$, which means the gravitational and the inertial masses are of equal value in rest. If the value of m_T is changing in time, then $m_V \neq m_T$. The potential (gravitational) mass coincides with the rest mass of the given particle and is stationary. What is changing along with the velocity (by the Lorentz transformation) is the kinetic (inertial) mass. We see that the equivalence of gravitational and inertial masses is one of the conditions to derive the Dirac equation, and this condition follows from (d1) to (d5), read from the conditions set up for the α and β coefficients.

(Condition (d4) makes possible to conclude a similar expression for the gradient of the ith component of the kinetic mass current in the form, which we can neglect in slowly changing systems, especially near to the rest.

$$(m_T - m_V)\frac{\partial \psi}{\partial x_i} = -\left(\frac{\partial m_T}{\partial x_i}\right)\psi$$

However, we cannot disregard it for we discuss a relativistic theory.)

Replacing α_i and β with appropriate practical multiplets and notations, Dirac introduced the matrices named after him.

In the presence of an arbitrary electromagnetic field with a scalar potential $\Phi = A_4$ and vector potential $\mathbf{A}$, we substitute $p_4 + (e_T/c)A_4$ for p_4, and $p_i + (e_V/c)A_i$ for p_i in the Hamiltonian for no field, where e_V and e_T denote the potential (Coulomb) and kinetic (Lorentz) charges. Remember that according to the assumption introduced in Section 3.5, there appear potential charges in the scalar field potential (A_4), which interacts solely with kinetic charges, and *vice versa*, there appear kinetic charges in the vector field potential ($\mathbf{A}$), which interacts solely with potential charges. Let us remember that e_T transforms with the velocity in a different way than m_T. In fact, it is not just the value of the charge of e_T, what changes at relativistic velocities, rather the charge density. Thus, similar to m_T, the density of e_T takes also three different values according to the spatial directions, like three components of a three-vector.

Introducing the earlier-deduced conditions in Eq. (11.8), the Dirac matrices (that follow from those conditions), and making the mentioned replacements to consider the effects of an electromagnetic field on our wave equation, we obtain:

$$\begin{aligned}&\left[-\left(p_4+\frac{e_T}{c}A_4\right)-\gamma_5\left(\boldsymbol{\sigma},\mathbf{p}+\frac{e_V}{c}\mathbf{A}\right)+\gamma_4 m_V c\right]\cdot\\&\cdot\left[\left(p_4+\frac{e_T}{c}A_4\right)-\gamma_5\left(\boldsymbol{\sigma},\mathbf{p}+\frac{e_V}{c}\mathbf{A}\right)+\gamma_4 m_T c\right]\psi=0\end{aligned}\tag{11.10}$$

(According to the convention, we replaced the ρ matrices applied in Dirac's original (1928) paper with the more widespread γ matrices, so that $\rho_1 = -\gamma_5$ and $\rho_3 = \gamma_4$, and also following the convention, we replace the original h in Dirac's equations with $\hbar$. To get a more easily comparable equation with the original, derived by Dirac – among other algebraic transformations – we make also the following identity replacement: $m_V m_T c^2 \equiv m_V^2 c^2 + m_V(m_T - m_V)c^2$. Also, we use – during the transformation of the wave equation – that the differential operators are ineffective on the stationary m_V and e_V.) We derive

$$\left\{\begin{array}{l}\left[\begin{array}{c}-\left(p_4+\frac{e_T}{c}A_4\right)^2+\left(\mathbf{p}+\frac{e_V}{c}\mathbf{A}\right)^2+m_V^2c^2+\\+\hbar\left(\boldsymbol{\sigma},\mathrm{rot}\left(\frac{e_V}{c}\mathbf{A}\right)\right)+i\hbar\gamma_5\left(\boldsymbol{\sigma},\mathrm{grad}\left(\frac{e_T}{c}A_4\right)+\frac{1}{c}\frac{\partial}{\partial t}\left(\frac{e_V}{c}\mathbf{A}\right)\right)\end{array}\right]+\\+\gamma_4\left[-\left(p_4+\frac{e_T}{c}A_4\right)+\gamma_5\left(\boldsymbol{\sigma},\mathbf{p}+\frac{e_V}{c}\mathbf{A}\right)+\gamma_4 m_V c\right](m_T-m_V)c\end{array}\right\}\psi=0\tag{11.11}$$

The first three terms in the first [ ] square bracket coincide with those in the relativistic wave equation for electromagnetic fields derived by Dirac (with the assumption $m_V = m_0$) with the only difference that we made a qualitative distinction between the potential (Coulomb) and kinetic (current or Lorentz) charges.

The fourth and fifth terms include $\mathrm{rot}\left(\frac{e_V}{c}\mathbf{A}\right) = e_V\mathbf{H}$, where $\mathbf{H}$ is the *magnetic vector* of the field; as well as the *electric vector* of the field in a modified form, where the potential and the kinetic charges are taken into account: $\mathrm{grad}\left(\frac{e_T}{c}A_4\right)+\frac{1}{c}\frac{\partial}{\partial t}\left(\frac{e_V}{c}\mathbf{A}\right) = e'\mathbf{E}$, where e' is a quantum mixture of e_V and e_T. The charges appear under the derivation because the value of e_T changes in relativistic covariant fields (for it is a function of its velocity in the given frame, cf., e.g. Achiezer and Berestetskii (1969, §22), and we are free to write e_V also under the time derivative because the derivative operator does not affect the potential charge e_V. More precisely, it is rather the density of e_T, which changes with its velocity. So, in the following, we replace the isotopic charges e_T with ρ_T and e_V with ρ_V in the formulas.

The expression in the first [ ] square bracket in (11.11) – with the mentioned alteration in the charges – coincides with the quadratic form of the Dirac equation.

Equation (11.11) differs from Dirac's result essentially in the last additional term:

$$-\gamma_4\left[-\left(p_4+\frac{\rho_T}{c}A_4\right)+\gamma_5\left(\boldsymbol{\sigma},\mathbf{p}+\frac{\rho_V}{c}\mathbf{A}\right)+\gamma_4 m_V c\right](m_T-m_V)c$$

This expression can be written by inserting the p_4 and $\mathbf{p}$ operators in the following:

$$-\gamma_4\left[-\left(\frac{i\hbar}{c}\frac{\partial}{\partial t}+\frac{\rho_T}{c}A_4\right)-\gamma_5\left(\boldsymbol{\sigma},i\hbar\ \mathrm{grad}-\frac{\rho_V}{c}\mathbf{A}\right)+\gamma_4 m_V c\right](m_T-m_V)c$$

The components in this term can be regarded as the additional energy of the interacting two massive, electrically charged particles due to their assumed additional degree of freedom (arbitrary positions in the IFCS field). They express the effect of the relativistic mass increase – the difference between the "dressed" and "bare" masses, that is, the "dress" in itself – on the electromagnetic field. We remind that this "dress" caused the precession of the velocity vector around the projection of the IFCS' third component and the rotation of the **D** field (cf., Section 10.3.3). The expression in this last square bracket [ ] coincides again with the Dirac wave equation, in its Hamiltonian form.

This last part of the equation gives an account on the *cross-interaction of the electromagnetic field and its two IFC with the two kinds of isotopic masses in quantum electrodynamics (QED)*. The state function ψ in this equation, unlike in the original Dirac equation, depends not only on the space–time co-ordinates and the spin but also on a two-valued variable that makes a distinction between the IFC.

In rest (when $m_{T=} m_V$, ρ_T=ρ_V), Eq. (11.11) coincides with the original Dirac equation. However, in relativistic covariant fields, the charges (densities) of both the gravitational and the electromagnetic fields will differ not only qualitatively, but also in their quantity, and we must take into account the last term. The appearance of this last term brings in the already acquainted (cf., Darvas, 2011, sections 3 and 3.2) inconvenient, but not unexpected, asymmetry in our theory that should be counteracted by the presumed HySy transformation between the states of the IFC.

The effects of the operators in the two [ ] square brackets in Eq. (11.11) must be equal:

$$\begin{bmatrix} -\left(p_4+\frac{\rho_T}{c}A_4\right)^2+\left(\mathbf{p}+\frac{\rho_V}{c}\mathbf{A}\right)^2+m_V^2c^2+\hbar\left(\boldsymbol{\sigma},\operatorname{rot}\left(\frac{\rho_V}{c}\mathbf{A}\right)\right)+ \\ +i\hbar\gamma_5\left(\boldsymbol{\sigma},\operatorname{grad}\left(\frac{\rho_T}{c}A_4\right)+\frac{1}{c}\frac{\partial}{\partial t}\left(\frac{\rho_V}{c}\mathbf{A}\right)\right) \end{bmatrix}\psi= \tag{11.12}$$
$$=\gamma_4\left[\left(p_4+\frac{\rho_T}{c}A_4\right)-\gamma_5\left(\boldsymbol{\sigma},\mathbf{p}+\frac{\rho_V}{c}\mathbf{A}\right)-\gamma_4 m_V c\right](m_T-m_V)c\psi$$

In the case of classical QED, the left side is equal to 0. The right side is 0, if m_T=m_V, that means, in a non-relativistic approximation. The effect of the operator in bracket { } on ψ in Eq. (11.11) will vanish as a result of the operators in the two square [ ] brackets together. If we demand that our mathematical derivations be in agreement with the time-proven Dirac equation, we must require that the effect of the operators in the first and the second square brackets on the wave function ψ be equal to 0 separately, according to the two sides of Eq. (11.12). Thus, our Eq. (11.11) separates into two equations.

The *first* equation

$$\begin{bmatrix} -\left(p_4+\frac{\rho_T}{c}A_4\right)^2+\left(\mathbf{p}+\frac{\rho_V}{c}\mathbf{A}\right)^2+m_V^2c^2+ \\ +\hbar\left(\boldsymbol{\sigma},\operatorname{rot}\left(\frac{\rho_V}{c}\mathbf{A}\right)\right)+i\hbar\gamma_5\left(\boldsymbol{\sigma},\operatorname{grad}\left(\frac{\rho_T}{c}A_4\right)+\frac{1}{c}\frac{\partial}{\partial t}\left(\frac{\rho_V}{c}\mathbf{A}\right)\right) \end{bmatrix}\psi=0 \tag{11.13}$$

will provide the solutions of the Dirac equation in the presence of potential and kinetic charges in an electromagnetic field. Note that there appears only the rest mass (m_V=m_0) of the particle. This equation differs from the original Dirac equation only in the presence of the two different isotopic electric charges.

The *second* equation

$$-\gamma_4\left[-\left(p_4+\frac{\rho_T}{c}A_4\right)+\gamma_5\left(\boldsymbol{\sigma},\mathbf{p}+\frac{\rho_V}{c}\mathbf{A}\right)+\gamma_4 m_V c\right](m_T-m_V)c\psi=0 \tag{11.14}$$

holds either in rest when quantitatively m_T=m_V, or when the value of the operator in the square bracket is 0.

The form of Eq. (11.11) guarantees that in boundary conditions our result coincides with the traditional. In a state close to rest, the second part vanishes and we get back to the well-known Dirac equation (11.13). In extreme relativistic situation, when $m_T \gg m_V$=m_0 (i.e. we can neglect the first component in (11.11)) we get Eq. (11.14), and can divide the full modified Dirac equation by ($m_T - m_V$). Eq. (11.14) can be written in a Schrödinger type form of a wave equation. The Dirac expression in the square bracket in Eq. (11.14) can be transformed into

$$i\hbar\frac{\partial}{\partial t}\psi=\left[-\rho_T A_4-\gamma_5(\boldsymbol{\sigma},i\hbar c\operatorname{grad}-\rho_V\mathbf{A})+\gamma_4 m_V c^2\right]\psi \tag{11.15}$$

where $-\rho_T A_4 - \gamma_5(\boldsymbol{\sigma}, i\hbar c\,\text{grad} - \rho_V \mathbf{A}) + \gamma_4 m_V c^2 = \mathbf{H}$ is the Hamiltonian of the system. There appears only the rest energy of the particles. However, due to the difference between ρ_T and ρ_V, this equation cannot be linearised in the four charge current components unless the IFCS invariance rotates the two isotopic electric charges of the electric field into each other in an IFCS gauge field (cf., Section 7.5). This equation does not reflect the effect of that gauge field, because the Dirac equation expresses the interaction of the two electrons in the electromagnetic field, more precisely the scalar Coulomb field with the electromagnetic vector field. In this semi-classical approach, we have not taken into account the interaction with the IFCS gauge field.

Concerning the described semi-classical approach, we should make a historical auxiliary remark. Later, Dirac (1951a) considered that the classical theories of the electromagnetic field are *approximate* and are valid only if the accelerations of the electrons are small. He stated that the earlier problems of QED resulted not in quantization, rather in the incompleteness of the classical theory of electrons, and one must try to improve on it. For this reason, he *proposed to replace the application of the Lorentz condition with a gauge theory*. He emphasised also the Hamiltonian approach instead of the Lagrangian one. He introduced a function λ (which was different from the quantity introduced by Feynman (1949) and got a current $j_\mu = -\lambda(\partial S/\partial x^\mu + A^*_\mu)$ where S was a gauge function attributed to A, and λ could be chosen to be an arbitrary infinitesimal at one instant of time, while its value at other times was then fixed by the conservation law $\partial j_\mu/\partial x_\mu = 0$. This method resulted in the conclusion that the theory (as expected) involves only the ratio e/m, not e and m separately. This theory (Dirac, 1951a) did not introduce the interaction of the electron with the electromagnetic field as a perturbation, like in the 1929–1932 Dirac–Fermi–Breit theories. The electron of that new theory could not be considered apart from its interaction with the electromagnetic field. As Dirac mentioned: "The theory of the present paper is put forward as a basis for a passage to a quantum theory of electrons. . . . one can hope that its correct solution will lead to the quantization of electric charge . . ." and " . . . questions of the interaction of the electron with itself no longer arise." Then, a further model by Dirac (1962, p. 64) provided a possible solution for eliminating the runaway motions of the electron.

Dirac's (1951a) paper was an attempt to exclude approximations by perturbation in either direction. It was in harmony with the aim of Bethe and Fermi (1932) to show the equivalence of the perturbations applied by Breit (1929, 1932) and Møller (1931). In this respect, Dirac's models were kin to the present attempt, in which, instead of a perturbation, we acknowledge the asymmetric roles of the interacting charged particles (as it can be read originally in Møller (1931)) and apply a gauge theory that has led us to a quantised theory. Certain ideas are borrowed here from Schroer (2011). The theory applied in this section to QED and having been proposed in a general form in Darvas (2011) eliminates the runaway motions of the electron too, although in an alternative way. We will return to the elimination of the runaway motions of the electron's wave function in Section 12.2.23.

We have to refer to another historical issue concerning an interaction with an IFCS gauge field. At the end of their already cited paper, Bethe and Fermi (1932, p. 306) showed that the formula introduced by Møller holds also when one of the interacting particles is in a bound state.[11.4] They consider also the option that the two interacting particles emit two quanta, but they reject it, because (for symmetry consideration for the momentums of the two quanta) they take into account only identical type quanta to be emitted and absorbed. (Although, the emission of one quantum painted another asymmetry in the picture, in which they aimed at eliminating the asymmetry caused by Møller's scattering matrix.) This conclusion by Bethe and Fermi is a result of their artificial symmetrisation of the potentials and does not arise in the theory set forth, among others, in this book.

Returning to our original train of thought, we can construct the Lagrangian of the interacting coupled two-electron system in the fields of each other from this Hamiltonian. Due to the two kinds of charges, this L differs from the usual form that appears in most textbooks. In principle, one can derive the non-linear charge–four-current from this L. The condition of linearization is that the gauge field, in which the charge densities ρ_T and ρ_V can substitute for each other, become invariant under an arbitrary gauge transformation. We will consider the interaction with a kinetic, concretely, IFCS gauge field in the following section.

11.3.3 Isotopic electric charges in the presence of a kinetic gauge field

Let us introduce the kinetic gauge field **D** similarly like in the general field-theoretical approach learned in Section 7.3. As we saw in Section 3 that the vierbein j_ν does not transform like a vector, we cannot expect this property of the components of **D** either. This **D** kinetic gauge field is associated (in this instance) with the electromagnetic field. Therefore, we extend the Dirac equation, discussed in Section 3, with the components of this **D** gauge field. For **D** is a kinetic field, all the four of its components interact with the potential electric (Coulomb) charge ρ_V (as it was assumed on the cross-interaction mechanism in Section 3). Thus, we start from the following extended form of equation (11.10):

$$\begin{aligned}&\left[-\left(p_4+\frac{\rho_T}{c}A_4+\frac{\rho_V}{c}D_4\right)-\gamma_5\left(\boldsymbol{\sigma},\mathbf{p}+\frac{\rho_V}{c}\mathbf{A}+\frac{\rho_V}{c}\mathbf{D}\right)+\gamma_4 m_V c\right]\cdot\\&\cdot\left[\left(p_4+\frac{\rho_T}{c}A_4+\frac{\rho_V}{c}D_4\right)-\gamma_5\left(\boldsymbol{\sigma},\mathbf{p}+\frac{\rho_V}{c}\mathbf{A}+\frac{\rho_V}{c}\mathbf{D}\right)+\gamma_4 m_T c\right]\psi=0\end{aligned} \tag{11.16}$$

11.4 For those who are interested in the problem of identity and non-identity of particles, as well as symmetrisation, more detailed analyses are recommended in Dieks ande Lubberdink (2011), and in Section 2.2 in this book).

here D_4 is the fourth component of $\mathbf{D}$, and $\mathbf{D}$ depends on the velocity components $D_{\dot{\mu}} = D(\dot{x}^{\mu})$. (Concerning the equivalence of this description and the space–time plus velocity-dependent description see Darvas, 2011; Bethe and Fermi, 1932.) Note that D_4, being a component of the kinetic gauge field, interacts with the potential electric charge, in contrast to the A_4 scalar potential of the electric field in the first ( ) brackets, and $\mathbf{D}$ in the second ( ) brackets in (11.16), which is here a three-component, vector-like quantity. Executing the multiplication, applying the same transformations as in Section 3, and considering that $p_4 = i\frac{\hbar}{c}\frac{\partial}{\partial t}$ and $p_i = -i\hbar\frac{\partial}{\partial x^i}$, as well as commutation of the components, one gets

$$\left\{\begin{array}{l} \left[-\left(p_4 + \frac{\rho_T}{c}A_4 + \frac{\rho_V}{c}D_4\right)^2 + \left(\mathbf{p} + \frac{\rho_V}{c}\mathbf{A} + \frac{\rho_V}{c}\mathbf{D}\right)^2 + m_V^2 c^2\right] + \\ + \hbar\left(\boldsymbol{\sigma}, \operatorname{rot}\left(\frac{\rho_V}{c}\mathbf{A}\right)\right) + i\hbar\gamma_5\left(\boldsymbol{\sigma}, \operatorname{grad}\left(\frac{\rho_T}{c}A_4\right) + \frac{1}{c}\frac{\partial}{\partial t}\left(\frac{\rho_V}{c}\mathbf{A}\right)\right) + \\ + \hbar\left(\boldsymbol{\sigma}, \operatorname{rot}\left(\frac{\rho_V}{c}\mathbf{D}\right)\right) + \hbar\left(\boldsymbol{\sigma}, \frac{\rho_V^2}{c^2}\left[D_j D_k - D_k D_j\right]\right) + \\ + i\hbar\gamma_5\left(\boldsymbol{\sigma}, \operatorname{grad}\left(\frac{\rho_V}{c}D_4\right) + \frac{1}{c}\frac{\partial}{\partial t}\left(\frac{\rho_V}{c}\mathbf{D}\right)\right) + \gamma_5\frac{\rho_V^2}{c^2}\left(\boldsymbol{\sigma}, D_4\mathbf{D} - \mathbf{D}D_4\right) + \\ + \gamma_4\left[-\left(p_4 + \frac{\rho_T}{c}A_4 + \frac{\rho_V}{c}D_4\right) + \gamma_5\left(\boldsymbol{\sigma}, \mathbf{p} + \frac{\rho_V}{c}\mathbf{A} + \frac{\rho_V}{c}\mathbf{D}\right) + \gamma_4 m_V c\right](m_T - m_V)c \end{array}\right\}\psi = 0 \tag{11.17}$$

Eq. (11.17) is the extended Dirac equation in the presence of isotopic electric charges and a kinetic gauge field **D**. There was considered that both the components of the EM vector potential **A**, and the elements of **D** commute with **σ**, the components of **A** commute with each other, but, for the elements of **D** do not compose a four-vector (in contrast to the components of **A**), we have no reason to assume that the elements of **D** would commute with each other. Thus, in the multiplication in (11.16) we considered the following equalities:

$$\begin{array}{l}\left(\boldsymbol{\sigma}, \mathbf{p} + \frac{\rho_V}{c}\mathbf{A} + \frac{\rho_V}{c}\mathbf{D}\right)^2 = \left(\mathbf{p} + \frac{\rho_V}{c}\mathbf{A} + \frac{\rho_V}{c}\mathbf{D}\right)^2 + \hbar\left(\boldsymbol{\sigma}, \operatorname{rot}\left(\frac{\rho_V}{c}\mathbf{A}\right)\right) + \\ + \hbar\left(\boldsymbol{\sigma}, \operatorname{rot}\left(\frac{\rho_V}{c}\mathbf{D}\right)\right) + \hbar\left(\boldsymbol{\sigma}, \frac{\rho_V^2}{c^2}\left(D_j D_k - D_k D_j\right)\right)\end{array}$$

and

$$\gamma_5\left(p_4+\frac{\rho_T}{c}A_4+\frac{\rho_V}{c}D_4\right)\left(\boldsymbol{\sigma},\mathbf{p}+\frac{\rho_V}{c}\mathbf{A}+\frac{\rho_V}{c}\mathbf{D}\right)-$$
$$-\gamma_5\left(\boldsymbol{\sigma},\mathbf{p}+\frac{\rho_V}{c}\mathbf{A}+\frac{\rho_V}{c}\mathbf{D}\right)\left(p_4+\frac{\rho_T}{c}A_4+\frac{\rho_V}{c}D_4\right)=$$
$$i\hbar\gamma_5\left(\boldsymbol{\sigma},\operatorname{grad}\left(\frac{\rho_T}{c}A_4\right)+\frac{1}{c}\frac{\partial}{\partial t}\left(\frac{\rho_V}{c}\mathbf{A}\right)\right)+$$
$$+i\hbar\gamma_5\left(\boldsymbol{\sigma},\operatorname{grad}\left(\frac{\rho_V}{c}D_4\right)+\frac{1}{c}\frac{\partial}{\partial t}\left(\frac{\rho_V}{c}\mathbf{D}\right)\right)+\gamma_5\frac{\rho_T^2}{c^2}(\boldsymbol{\sigma},D_4\mathbf{D}-\mathbf{D}D_4)$$

This extended form (11.17) of the Dirac equation can be formulated in a more elegant form with the help of the HySy group (cf., Section 11.3.4).

11.3.4 Hypersymmetry of the extended Dirac equation

In the course to find an algebraic form of the so extended Dirac equation, first remind that we already showed that Δ (IFCS) is a two-valued spin-like property that appears in the presence of a kinetic gauge field at high energies. A qualitative distinction was made between the two (isotopic) kinds of field-charges.[11.5]

We apply the τ-algebra for the Dirac equation as we did it for the gravitational stress–energy tensor in the Section (11.2). The original Dirac equation was invariant under the Lorentz transformation, however, this property did not ensure sufficiency for the covariance of the equation under strongly relativistic conditions (this will be discussed in Section 11.3.6). In the previous section, we introduced strongly relativistic conditions in the source of the Dirac equation. This extension will be discussed in the Section 11.3.5 in the light of the HySy algebra.

A particle can occupy either the one or the other IFC state. The wave function of a given particle may be in a potential state with amplitude ψ_V or a kinetic state with amplitude ψ_T.[11.6,11.7] We repeat that the wave function of a single physical particle, in case of the electromagnetic interaction too, is an entanglement of the probabilities

11.5 Similar distinction was discussed for the Lagrangians of a system in Dirac (1928). They follow a Lagrangian-based way to find a Noether symmetry, like the author derived in Section 7.3 and Darvas (2012d), although their derivation of the results seems more complicated, and less general (i.e. holds only for the gravitational interaction).

11.6 On partition between the potential and kinetic energies, see 't Hooft (2002).

11.7 As shown in Section 10.1, according to one of the interpretations of the IFCS theory, the energy of a single particle at a given moment is considered to be ideally concentrated either in the scalar or in the kinetic part of its Hamiltonian (Darvas and Farkas, 2008). The corresponding state functions belonging to these two idealised reduced Hamiltonians are denoted by ψ_V and ψ_T. They characterise the s.c. *potential* and *kinetic* states of the particle. A single particle takes one of these two states at certain *probabilities*. According to this interpretation the particle can change its state (or oscillate) between the fully potential and the fully kinetic states. The observable state function can being in the potential or the kinetic state at a given moment: $\psi = \begin{pmatrix} \psi_T \\ \psi_V \end{pmatrix}$, where ψ_T is a three-component column (cf., footnote 11.7). We denoted the eigenfunctions ψ_T and ψ_V that belong to the two eigenvalues of the operator τ by $\varphi_+^{(\tau)}$ and $\varphi_-^{(\tau)}$, respectively. There belong two opposite IFCS (Δ) positions to the sources of the two states. During the transition from ψ_V to ψ_T and back, the source of the field, that is, the respective field-charge, needs to change its Δ state. The operator $\boldsymbol{\tau}$, whose matrix algebra we have introduced in Section 5.2 in this book, affects those Δ states.

As shown there, χ and ϑ are the eigenfunctions of the Δ IFCS's $\boldsymbol{\tau}$ operators. (As mentioned, the corresponding eigenfunctions were denoted also by $\varphi_+^{(\tau)}$ and$\varphi_-^{(\tau)}$). The eigenvalue equations are

$$\tau\varphi_+^{(\tau)} = \tfrac{1}{2}\varphi_+^{(\tau)}\,(\text{or } \tau\chi = \tfrac{1}{2}\chi)\text{ and } \tau\varphi_-^{(\tau)} = -\tfrac{1}{2}\varphi_-^{(\tau)}\,(\text{or } \tau\vartheta = -\tfrac{1}{2}\vartheta),$$

respectively. The eigenvalues of Δ can take $\pm(1/2)$ as we saw it also in Section 11.2.2 devoted to gravity. The operator $\boldsymbol{\tau}$ refers to the transformation matrices of the *HySy group* discussed in Section 5.2.5 that rotate, for example, the IFCS, in the introduced kinetic gauge field. Since the operators responsible for the rotation in the kinetic gauge field do not affect the space–time-dependent and the spin-dependent components in the field equations, and *vice versa*, the particle's full state function ψ can be separated according to the following sum: $\psi = \psi'\varphi_+^{(\tau)} + \psi''\varphi_-^{(\tau)}$, where ψ' and ψ" denote the state functions affected by the space–time-dependent and the spin-dependent operators. The ψ state functions in (11.18), (11.19), and (11.21) are functions of the x_ν space–time co-ordinates, and of the $r = \pm(1/2)$ and $s = \pm(1/2)$ parameters (according to the set of eigenvalues of the $\boldsymbol{\rho}$ and $\boldsymbol{\sigma}$ operators), as well as – thanks to the extension of the Dirac equation in Section 11.3.3 – of the $\Delta = \pm(1/2)$ parameters (according to the set of eigenvalues of the $\boldsymbol{\tau}$ operator). Let us denote the respective eigenfunctions belonging to the eigenvalues of the r, s, and Δ parameters in the state function ψ in the following way: $\varphi_+^{(\rho)}, \varphi_-^{(\rho)}, \varphi_+^{(\sigma)}, \varphi_-^{(\sigma)}$, $\varphi_+^{(\tau)}, \varphi_-^{(\tau)}$, according to the $\pm(1/2)$ eigenvalues of the $\boldsymbol{\rho}$, $\boldsymbol{\sigma}$, and $\boldsymbol{\tau}$ operators. The differential operators act only on $\psi(x_\nu)$, the $\boldsymbol{\sigma}$ operators on the eigenfunctions of σ, the $\boldsymbol{\rho}$ operators on the eigenfunctions of ρ, and the $\boldsymbol{\tau}$ operators on the eigenfunctions of Δ. (To explain this latter notation, please note, we cannot denote the eigenfunctions, which the $\boldsymbol{\tau}$ operators act on, by t, first, because t denotes time; second, consistency is required with the notations used for the same purpose in former publications.)

be characterised by the entanglement of these two probability states and denoted by ψ. The IFCS theory describes transition between the two states.

Thus, the $\psi = \psi(x_v, r, s, \Delta)$state function can be dissociated to the sum of the following eight products according to the eigenfunctions of the matrix operators:

$$\begin{aligned}
\psi &= \psi' \varphi_+^{(\tau)} + \psi'' \varphi_-^{(\tau)} = \\
&\begin{bmatrix} \psi_1'(x_v)\varphi_+^{(\rho)}\varphi_+^{(\sigma)} + \psi_2'(x_v)\varphi_-^{(\rho)}\varphi_+^{(\sigma)} + \\ + \psi_3'(x_v)\varphi_+^{(\rho)}\varphi_-^{(\sigma)} + \psi_4'(x_v)\varphi_-^{(\rho)}\varphi_-^{(\sigma)} \end{bmatrix} \varphi_+^{(\tau)} + \\
&+ \begin{bmatrix} \psi_1''(x_v)\varphi_+^{(\rho)}\varphi_+^{(\sigma)} + \psi_2''(x_v)\varphi_-^{(\rho)}\varphi_+^{(\sigma)} + \\ + \psi_3''(x_v)\varphi_+^{(\rho)}\varphi_-^{(\sigma)} + \psi_4''(x_v)\varphi_-^{(\rho)}\varphi_-^{(\sigma)} \end{bmatrix} \varphi_-^{(\tau)} = \\
&= \psi_1'(x_v)\varphi_+^{(\rho)}\varphi_+^{(\sigma)}\varphi_+^{(\tau)} + \psi_1''(x_v)\varphi_+^{(\rho)}\varphi_+^{(\sigma)}\varphi_-^{(\tau)} + \\
&+ \psi_2'(x_v)\varphi_-^{(\rho)}\varphi_+^{(\sigma)}\varphi_+^{(\tau)} + \psi_2''(x_v)\varphi_-^{(\rho)}\varphi_+^{(\sigma)}\varphi_-^{(\tau)} + \\
&+ \psi_3'(x_v)\varphi_+^{(\rho)}\varphi_-^{(\sigma)}\varphi_+^{(\tau)} + \psi_3''(x_v)\varphi_+^{(\rho)}\varphi_-^{(\sigma)}\varphi_-^{(\tau)} + \\
&+ \psi_4'(x_v)\varphi_-^{(\rho)}\varphi_-^{(\sigma)}\varphi_+^{(\tau)} + \psi_4''(x_v)\varphi_-^{(\rho)}\varphi_-^{(\sigma)}\varphi_-^{(\tau)}
\end{aligned} \tag{11.18}$$

This involves also that the solutions of the extended Dirac equation's ψ state function can be separated into eight space–time-dependent functions. Opposite to the four solutions of the original Dirac equation, this extended equation has eight solutions, in accordance with the additional bivariant opposite positions of the IFCS. Important to notice: the coincidence of the set of matrices for the τ- and the T-algebras (cf., Section 5.3.1) guarantees that the number of solutions is *doubled*, thanks to the IFCS invariance (according to the two new degrees of freedom brought in by the two possible positions of the IFCS),[11.8] compared with the *quadruple* solutions of the original Dirac equation due to the non-coincidence of the $\boldsymbol{\rho}$ and $\boldsymbol{\sigma}$ operators (Dirac, 1928, 1929). It can be exemplified so that, according to the two isotopic states of the field-charges, two separated solutions are assigned to each of the four known solutions of the Dirac equation.

11.3.5 Application of the HySy algebra for the extended Dirac equation

To apply the HySy algebra, let us start from the source of the extended Dirac equation[11.9] as derived in Section 11.3.3 Eq. (11.16) (Darvas, 2013a) in the presence

11.8 Jentschura bases (cf., Jentschura, 2013; Noble and Jentchura, 2015a, b) his discussion of the multiple solutions of the Dirac equation to the considerations discussed in Section 5.3.1, items (1)–(4).
11.9 Modifications of the Dirac equation among specific conditions, for example, Darvas (2014) presented also new solutions for it. Apart from giving quaternionic form to potentials, the two different new solutions provide a remarkable demonstration that non-traditional approaches to the electrically charged particle interactions are an actual problem in contemporary physics. Moreover, Podolsky discussed its application for half-vectors, showing independence of the choice of n-beins, already in 1931.

of isotopic electric charge densities (ρ_V, ρ_T) and isotopic gravitational charges (m_V, m_T), on the one hand, as well as a kinetic gauge field (**D**) on the other hand:

$$\begin{aligned}&\left[-\left(p_0+\frac{\rho_T}{c}A_4+\frac{\rho_V}{c}D_4\right)-\gamma_5\left(\boldsymbol{\sigma},\mathbf{p}+\frac{\rho_V}{c}\mathbf{A}+\frac{\rho_V}{c}\mathbf{D}\right)+\gamma_4 m_V c\right]\cdot\\&\cdot\left[\left(p_0+\frac{\rho_T}{c}A_4+\frac{\rho_V}{c}D_4\right)-\gamma_5\left(\boldsymbol{\sigma},\mathbf{p}+\frac{\rho_V}{c}\mathbf{A}+\frac{\rho_V}{c}\mathbf{D}\right)+\gamma_4 m_T c\right]\psi=0\end{aligned} \tag{11.19}$$

Note again that $\gamma_4=\rho_3$, $\gamma_5=-\rho_1$, and let us apply the τ algebra for the eigenvectors of the operators affecting the IFC. For simplicity, let us introduce the following expressions for the generalised momenta (following Dirac's original notations, extended with **D**):

$$p_4'=-i\left(p_0+\frac{\rho_T}{c}A_4+\frac{\rho_V}{c}D_4\right);\ \mathbf{p}'=\mathbf{p}+\frac{\rho_V}{c}\mathbf{A}+\frac{\rho_V}{c}\mathbf{D}.$$

As shown in Section 5.2.3, the IFC – in our case both the electric and the gravitational – are rotated by the same transformation matrices in the *IFCS* field.[11.10, 11.11]

Let us investigate the operators affecting ψ' and ψ" in (11.18):$\psi = \psi'\varphi_+^{(\tau)}+\psi''\varphi_-^{(\tau)}$, separately. We avoid details of the calculations (one can find them in Darvas (2015b)):

$$\begin{aligned}&\left[i\tau_3\rho_3 p_4'+i\mathbf{E}\,\rho_2(\boldsymbol{\sigma},\mathbf{p}')+\mathbf{E}m_T c\right]\psi'\chi=\\&=\left[i\delta_4\gamma_4 p_4'+i\delta(\boldsymbol{\gamma},\mathbf{p}')+\mathbf{E}m_T c\right]\psi'\chi\end{aligned} \tag{11.20'}$$

$$\begin{aligned}&\left[i\tau_3\rho_3 p_4'+i\mathbf{E}\,\rho_2(\boldsymbol{\sigma},\mathbf{p}')+\mathbf{E}m_T c\right]\psi''\vartheta=\\&=\left[i\delta_4\gamma_4 p_4'+i\delta(\boldsymbol{\gamma},\mathbf{p}')+\mathbf{E}m_T c\right]\psi''\vartheta\end{aligned} \tag{11.20''}$$

Note that **E** denotes here the unit matrix of the τ algebra, do not confuse it with the electric field strength. After rearrangement, the source of the extended Dirac equation (11.19)) can be written in the following simple form:

$$\left[i\delta_4\gamma_4 p_4'+i\delta(\boldsymbol{\gamma},\mathbf{p}')+\mathbf{E}\ m_V c\right]\cdot\left[i\delta_4\gamma_4 p_4'+i\delta(\boldsymbol{\gamma},\mathbf{p}')+\mathbf{E}\ m_T c\right]\psi=0 \tag{11.21}$$

11.10 An approach, similar to our one, was followed by Deriglazov and Nersessian (2014). They found – when looking for passages from classical to quantum theory using SO(3.2)-algebra – that quantization of the model led to massive Dirac equation.

11.11 The role of the mass term in the Dirac equation, confirming the consistency of quantum mechanics with general relativity, was discussed in detail by Goenner (2013), which confirmed our assumption to take into account the difference between the gravitational and inertial masses in the Dirac equation. Jentschura showed the possibility to bring the gravitationally coupled Dirac equation to a form where it can easily be unified with the electromagnetic coupling as it is commonly used in modern particle physics calculations.

We introduce the following notations:

$$\Gamma_1 = \delta_1\gamma_1 = \tau_2 = \begin{bmatrix} 0 & 0 & 0 & -i \\ 0 & 0 & 0 & -i \\ 0 & 0 & 0 & -i \\ i & 0 & 0 & 0 \end{bmatrix}; \quad \Gamma_2 = \delta_2\gamma_2 = -\tau_1 = \begin{bmatrix} 0 & 0 & 0 & -1 \\ 0 & 0 & 0 & -1 \\ 0 & 0 & 0 & -1 \\ -1 & 0 & 0 & 0 \end{bmatrix};$$
$$\Gamma_3 = \delta_3\gamma_3 = \tau_3\rho_2 = \begin{bmatrix} 0 & 0 & -i & 0 \\ 0 & 0 & -i & 0 \\ 0 & 0 & -i & 0 \\ 0 & -i & 0 & 0 \end{bmatrix}; \quad \Gamma_4 = \delta_4\gamma_4 = \mathbf{E} = \begin{bmatrix} 1 & 0 & 0 & 0 \\ 1 & 0 & 0 & 0 \\ 1 & 0 & 0 & 0 \\ 0 & 0 & 0 & 1 \end{bmatrix} \tag{11.22}$$

Note also that $\Gamma_3 = -i\mathbf{E}\,\rho_1 = i\mathbf{E}\,\gamma_5$.

Now, source (11.21) of the extended Dirac equation can be written with these notations in the following *more simple* form:

$$\left[i\Gamma_\mu p'_\mu + \mathbf{E}\ m_V c\right] \cdot \left[i\Gamma_\nu p'_\nu + \mathbf{E}\ m_T c\right]\psi = 0 \tag{11.23}$$

The latter is the *extended form of the Dirac equation's source in the presence of IFC with the application of the HySy tau-algebra.* All the rest can be calculated on the bases of Darvas, 2011, 2013b, 2014 as well as Sections 11.3.8–11.3.12. The form of the source equation justifies the duplication of the solutions, indicated earlier following Eq. (11.18). One can observe that the uppercase Greek *Γ* plays a similar role like the lowercase Greek Dirac *γ* did in the original equation. They represent the combined transformation matrices of the Dirac algebra and HySy.

Recall again that the transformation of the field-charges in the abstract IFCS field is independent of the given physical interaction (gravitational, electromagnetic, weak, strong); the same transformation rotates the field-charges of all fundamental interactions between their scalar (potential) and vector (kinetic) states appearing in the respective parts of their Hamiltonian. The latter remark holds, of course, also for both the field-charges of the electromagnetic and the gravitational fields. The chiral character of the IFCS transformation gets special significance in the case of the electro-weak interaction.

11.3.6 Invariance of the extended Dirac equation

The (11.23) form $\left[i\Gamma_\mu p'_\mu + \mathbf{E}\ m_V c\right] \cdot \left[i\Gamma_\nu p'_\nu + \mathbf{E}\ m_T c\right]\psi = 0$ of the extended Dirac equation's source demonstrates its covariant character. Two consequences, in comparison with the original Dirac equation, are apparent. First, the appearance of the opposite IFC in the p_ν' generalised momentum and in the mass terms destroy the

Lorentz invariance of the equation. Second, there appears a new invariance, represented by the δ [or in other notation (11.22) by the τ] operators that rotates the IFC in an additional abstract field. The covariance of the extended Dirac equation has resulted in the convolution of these two invariances. The extended Dirac equation is invariant under the combined transformation.

This combined (Lorentz $\otimes$ IFCS) means that ($SO^{+}(3,1) \otimes SU(2)$) invariance of the equation allows again to reinterpret our earlier imagination on the invariances of physical equations. It was a long-lasting paradigm in physics that equations of the physical interactions must be subject of invariance under the Lorentz transformation, and this was not only a necessary but also a sufficient condition. Now, we see this is not the case. Nevertheless, as we premised, no physical principle stated that Lorentz-invariance is a sufficient condition demanded from the physical equations. There exist other invariances that may appear together and combined with Lorentz's, especially among particular conditions.

11.3.7 Mechanism of the Δ exchange in electromagnetic interaction

The discussed interaction mechanism model (cf., Section 10.1) fits best to the electromagnetic interaction. Similar to the mass, two (i.e. potential and kinetic) states of the electric charges can be interpreted, and they correspond to the Coulomb charge and to the components of the charge current (Lorentz charges). There exists a similar equivalence principle between them, like between the two states of masses. Since the electromagnetic theory is subject to invariance under $U(1)$ group's symmetry transformations,

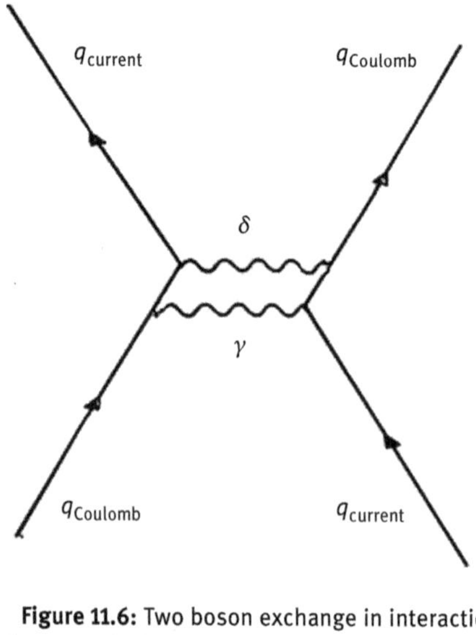

Figure 11.6: Two boson exchange in interaction between the field-charges of the electromagnetic field. (δ denotes the dion, γ denotes the photon.)

there belongs to it a single gauge quantum (γ). It mediates the interaction between particles with potential (Coulomb) state and kinetic (Lorentz) state charges. δ_{EM} is responsible to switch both actors to occupy the opposite Δ state after the interaction. The process, with the exchange of the two bosons, continues permanently.

This model corresponds to the asymmetric model described by Møller (1931). Taking into consideration the difference between the potential (ψ_V) and kinetic (ψ_T) states of $\urcorner$ (where, in this concrete case, $\urcorner$ denotes the electric charge), one does not need the artificial symmetrisation (cf., Section 3.1) introduced by Bethe and Fermi (1932), because the considered asymmetry is compensated by the Δ invariance mediated by δ_{EM}.

11.3.8 Modified Dirac equation in the presence of isotopic electric charges and a kinetic gauge field

Equation (11.17) can be written in the following form: $[W + W^A + W^D - H(m_T - m_V)c]\psi = 0$, where W refers to the first line of (11.17), W^A to the second line, W^D to the third and fourth lines, and $H(m_T - m_V)c$ to the fifth line of Eq. (11.17). They correspond to the contributing fields, respectively.

11.3.8.1 Coincidence with the classical Dirac equation in boundary case, when no kinetic field is present

The first line of the operator in Eq. (11.17), W, expresses the first three elements of the classical Dirac equation, with the modifications that it contains (a) the isotopic electric charges and (b) the kinetic vector potential $\mathbf{D}$ of the considered kinetic field.

11.3.8.2 The magnetic and the electric moments

The two elements in W^A – considering the isotopic electric charges – coincide with the two elements of the classical Dirac equation as discussed in Section 11.3.2, and yield the *magnetic* and the *electric moments* of the electron interacting with the electromagnetic field, respectively.

11.3.8.3 The magneto-kinetic and electro-kinetic moments

The essential difference, compared to Eq. (11.11) of the semi-classical QED model discussed in Section 11.3.2, is in W^D expressed in lines 3 and 4 of the Eq. (11.17). The expression

$$\begin{aligned}&\hbar\left(\boldsymbol{\sigma}, \mathrm{rot}\left(\frac{\rho_V}{c}\mathbf{D}\right)\right) + \hbar\left(\boldsymbol{\sigma}, \frac{\rho_V^2}{c^2}\left[D_j D_k - D_k D_j\right]\right) + \\ &+ i\hbar\gamma_5\left(\boldsymbol{\sigma}, \mathrm{grad}\left(\frac{\rho_V}{c}D_4\right) + \frac{1}{c}\frac{\partial}{\partial t}\left(\frac{\rho_V}{c}\mathbf{D}\right)\right) + \gamma_5\frac{\rho_V^2}{c^2}\left(\boldsymbol{\sigma}, D_4\mathbf{D} - \mathbf{D}D_4\right)\end{aligned} \tag{11.24}$$

provides a *kinetic moment* of the **D** field. Introducing the commutator of **D**, one can write the following:

$$\begin{aligned}&\hbar\left(\boldsymbol{\sigma},\operatorname{rot}\left(\frac{\rho_V}{c}\mathbf{D}\right)\right)+ig\hbar\frac{\rho_V^2}{c^2}(\boldsymbol{\sigma},C^i_{jk}D_jD_k)+\\&+i\hbar\gamma_5\left(\boldsymbol{\sigma},\operatorname{grad}\left(\frac{\rho_V}{c}D_4\right)+\frac{1}{c}\frac{\partial}{\partial t}\left(\frac{\rho_V}{c}\mathbf{D}\right)\right)+\gamma_5\frac{\rho_V^2}{c^2}(\boldsymbol{\sigma},D_4\mathbf{D}-\mathbf{D}D_4)\end{aligned}\tag{11.24a}$$

or

$$\begin{aligned}&\left(\frac{\hbar}{c}\boldsymbol{\sigma},\operatorname{rot}\ (\rho_V\mathbf{D})\right)+\left(\frac{\hbar}{c}\boldsymbol{\sigma},igC^i_{jk}\frac{\rho_V^2}{c}D_jD_k\right)+\\&+i\gamma_5\left[\left(\frac{\hbar}{c}\boldsymbol{\sigma},\operatorname{grad}\ (\rho_VD_4)+\frac{1}{c^2}\frac{\partial}{\partial t}(\rho_V\mathbf{D})\right)+\left(\frac{\hbar}{c}\boldsymbol{\sigma},\frac{\rho_V^2}{\hbar c}(\mathbf{D}D_4-D_4\mathbf{D})\right)\right]\end{aligned}\tag{11.24b}$$

Here C^i_{jk} are the structure constants appearing with the multiplication of D_is, and g is the coupling constant for the electromagnetic interaction. C^i_{jk} are the coefficients in the commutation rule of the generators (transformation matrices) of the group of HySy in the kinetic (IFC) field, as we saw in Section 7.3. Since this field is subject of an $SU(2)$ isomorphic symmetry, there are three C^i_{jk} structure constants. This commutation term does not appear in W^A, because the **A** vector potential of the EM field composes a vector and the derivatives of **A** commute with each other as vectors.

Since the derivatives of $D_{\dot{\mu}}=D(\dot{x}^\mu)$appearing in Eq. (11.17) are derived by the space–time co-ordinates and $D_{\dot{\mu}}$ depends on $\dot{x}^\mu$, all derivatives of $D_{\dot{\mu}}$ must be interpreted as $\frac{\partial D_{\dot{\mu}}}{\partial x^\nu}=\frac{\partial D_{\dot{\mu}}}{\partial \dot{x}^\rho}\frac{\partial \dot{x}^\rho}{\partial x^\nu}=D_{\dot{\mu}\dot{\rho}}\Lambda^{\dot{\rho}}_\nu=D_{\dot{\mu}\dot{\rho}}\dot{x}^\rho_\nu$ (where μ, ν, ρ=1, . . . , 4).

For simplicity, let us assume that **D** depends only on the linear combinations of the first derivatives and multiplications of the velocity, on the velocity itself, as well as on $\kappa=1/\sqrt{1-(\dot{x}_i^2/c^2)}$ and a constant. So

$$D(\dot{x}^\mu)=\alpha\frac{\partial\dot{x}^\mu}{\partial\dot{x}^\rho}\frac{\partial\dot{x}^\rho}{\partial x^\nu}+\beta\dot{x}_i\dot{x}_j+\gamma\dot{x}_i+\delta\kappa+\varepsilon$$

where α, β, γ, δ, and ε are coefficients, not depending on the actual relative velocity of the interacting charges. In this plausible, but quite not the most general case, the commutator of D_j and D_k is not identically 0. However, all the three elements of D_i (i=1, 2, 3) commute with D_4.

In the general case, (11.24b) can be written as $((\hbar/c)\boldsymbol{\sigma},\mathbf{M}^D)+i\gamma_5((\hbar/c)\boldsymbol{\sigma},\mathbf{N}^D)$, where

$$\mathbf{M}^D=\operatorname{rot}\,\rho_V\mathbf{D}+igC^i_{jk}\frac{\rho_V^2}{c}D_jD_k\ \text{and}$$

$$\mathbf{N}^D=\operatorname{grad}\,\rho_VD_4+\frac{1}{c^2}\frac{\partial}{\partial t}\rho_V\mathbf{D}+\frac{\rho_V^2}{\hbar c}(\mathbf{D}D_4-D_4\mathbf{D})\tag{11.25}$$

With the mentioned conditions, the third term in $\mathbf{N}^D$ vanishes. Note that in the expressions of $\mathbf{M}^D$ and $\mathbf{N}^D$, there appear only the potential (Coulomb) charge densities.

This is natural because the velocity dependence is considered in the kinetic gauge potential **D**, which these charges interact with. Since ρ_V does not depend either on space–time co-ordinates or on the actual velocity, it is not subject of derivation:

$$\mathbf{M}^D = \rho_V \operatorname{rot} \mathbf{D} + igC^i_{jk}\frac{\rho_V^2}{c}D_jD_k \quad \text{and} \quad \mathbf{N}^D = \rho_V \operatorname{grad} D_4 + \frac{\rho_V}{c}\frac{\partial}{\partial t}\mathbf{D} + \frac{\rho_V^2}{\hbar c}(\mathbf{D}D_4 - D_4\mathbf{D})$$

The kinetic moment is an additional, new quantity in the isotopic electric charge theory compared to the classical Dirac theory. The two kinetic moments determine the isotopic electric charge spin Δ_{el}. According to Section 8.2, the IFCS (including also the isotopic electric charge spin) is a conserved quantity, so it must commute with the Hamiltonian.

11.3.8.3.1 The magneto-kinetic moment

The first term in *Eq. (11.24)*, $\mathbf{M}^D$ (with an $(m_T - m_V)$ divider) can be considered as a "magneto-kinetic" additional energy of the electric charge due to its additional degree of freedom assigned to it by the interaction with the kinetic field. For D_μ does not behave as a vector, its derivatives include an additional, gauge term, where the derivation of the extended Dirac equation (*11.17*) provided automatically in the form of the third term (W^D) in expression (11.17). So, this second term of $\mathbf{M}^D$ forms part of the "magneto-kinetic" momentum of the field (cf., in the first line of *Eq. (11.24)*).

The full magnetic moment of the interaction (with a mass-dimension divider), in the presence of the kinetic gauge field, will be

$$\left(\frac{\hbar}{c}\boldsymbol{\sigma}, \mathbf{M}^{\mathrm{FULL}}\right) = \left(\frac{\hbar}{c}\boldsymbol{\sigma}, \rho_V \operatorname{rot}(\mathbf{A}+\mathbf{D}) + igC^i_{jk}\frac{\rho_V^2}{c}D_jD_k\right)$$

11.3.8.3.2 The electro-kinetic moment

The third term in Eq. (11.24) is similar to the expression obtained for the electric moment of the EM field in line 2 (W^A) of (11.17), extended also with a gauge term. It can be considered similarly, like in Dirac's theory, as an "electro-kinetic" additional energy of the electron. Also, similar to the $\hbar\gamma_5\left(\boldsymbol{\sigma}, \operatorname{grad}\left(\frac{\rho_T}{c}A_4\right) + \frac{1}{c}\frac{\partial}{\partial t}\left(\frac{\rho_V}{c}\mathbf{A}\right)\right)$ electric moment in W^A, the "electro-kinetic" moment

$$i\gamma_5\left[\left(\frac{\hbar}{c}\boldsymbol{\sigma}, \rho_V \operatorname{grad} D_4 + \frac{\rho_V}{c}\frac{\partial}{\partial t}\mathbf{D}\right) + \left(\frac{\hbar}{c}\boldsymbol{\sigma}, \frac{\rho_V^2}{\hbar c}(\mathbf{D}D_4 - D_4\mathbf{D})\right)\right]$$

is apparently imaginary too. However, this is only an appearance. Dirac observed the following: "It is doubtful whether the electric moment has any physical meaning" as a result that the multiplication, due to which the imaginary term appeared, was an "artificial involvement" in the equation. Note, that while σ_1 and σ_3 are real, σ_2 is imaginary, so $i\sigma_2$ is real, and only $i\sigma_1$ and $i\sigma_3$ are imaginary. (At the same time, the second component in the magnetic moment is also imaginary, due to the imaginary character of σ_2.) In contrast to the original Dirac equation in the case of the extended Dirac equation, the multiplier of the three-vector $\boldsymbol{\sigma}$ (or $\gamma_5\boldsymbol{\sigma}$) is a sum, which includes the components of the velocity-dependent field. In the presence of the kinetic field **D**, one can choose the co-ordinate system fitted to the electron's velocity arrow so that the multiplier of the imaginary σ_2 be non-zero. Then the expression will yield *a real term* for the sum of the electric and the electro-kinetic energy, while there will be left *two imaginary terms* for the multiplication by σ_1 and σ_3, whose sum should be made equal to 0 and provide a constraint for the energy of the interaction. Note that these two expressions are not fully imaginary, since D_4, depending on the fourth component of the velocity, is imaginary itself. On the other side, the multiplier of σ_2 contains an imaginary component, too ($\text{grad}_2 D_4$). With these conditions, *one can eliminate the imaginary terms* in (11.17) and *give physical meaning to the electric and the electro-kinetic moments*:

$$
\begin{aligned}
& i\hbar\gamma_5\left(\sigma_1, \text{grad}_1\left(\frac{\rho_T}{c}A_4\right)+\frac{1}{c}\frac{\partial}{\partial t}\left(\frac{\rho_V}{c}A_1\right)\right)+\\
& +i\hbar\gamma_5\left(\sigma_1, \text{grad}_1\left(\frac{\rho_V}{c}D_4\right)+\frac{1}{c}\frac{\partial}{\partial t}\left(\frac{\rho_V}{c}D_1\right)\right)+\\
& +\gamma_5\frac{\rho_V^2}{\hbar c^2}(\sigma_1, D_4D_1-D_1D_4)=0\\
& i\hbar\gamma_5\left(\sigma_3, \text{grad}_3\left(\frac{\rho_T}{c}A_4\right)+\frac{1}{c}\frac{\partial}{\partial t}\left(\frac{\rho_V}{c}A_3\right)\right)+\\
& +i\hbar\gamma_5\left(\sigma_3, \text{grad}_3\left(\frac{\rho_V}{c}D_4\right)+\frac{1}{c}\frac{\partial}{\partial t}\left(\frac{\rho_V}{c}D_3\right)\right)+\\
& +\gamma_5\frac{\rho_V^2}{\hbar c^2}(\sigma_3, D_4D_3-D_3D_4)=0
\end{aligned}
$$

The two conditions of the earlier equation can be satisfied if both the imaginary and the real parts are equal to 0 separately

$$
\begin{aligned}
& i\hbar\left[\text{grad}_1\ (\rho_T A_4)+\frac{\rho_V}{c}\frac{\partial}{\partial t}(A_1+D_1)\right]+\frac{\rho_V^2}{\hbar c}(D_4D_1-D_1D_4)=0\\
& \text{grad}_1 D_4=0
\end{aligned}
$$

and

$$
\begin{aligned}
& i\hbar\left[\text{grad}_3(\rho_T A_4)+\frac{\rho_V}{c}\frac{\partial}{\partial t}(A_3+D_3)\right]+\frac{\rho_V^2}{\hbar c}(D_4D_3-D_3D_4)=0\\
& \text{grad}_3 D_4=0
\end{aligned}
$$

We add from among the multipliers of σ_2:

$$\mathrm{grad}_2 D_4 = 0$$

Provided that the components of the **A** vector potential and the value of the interacting charges are known, we obtained a set of differential equations to determine the components of D_μ, at an actual value of the velocity (cf., Section 10.3.3), arrowed parallel to σ_2, in each space–time point.

The reference frame, in which we calculated the constraints for **D**, rotates together with the kinetic charge density ρ_T (cf., Section 11.3.2). This choice provided a restriction for the interacting system, while at the same time, it made us possible to calculate the exact forms of the components of D_μ in the given reference frame.

Considering also the assumption formulated in Section 11.3.8.3, the set of differential equations reduces to the following:

$$\begin{aligned}
&\mathrm{grad}_1(\rho_T A_4) + \frac{\rho_V}{c}\frac{\partial}{\partial t}(A_1 + D_1) = 0; \quad && \mathrm{grad}_1 D_4 = 0 \\
& && \mathrm{grad}_2 D_4 = 0 \\
&\mathrm{grad}_3(\rho_T A_4) + \frac{\rho_V}{c}\frac{\partial}{\partial t}(A_3 + D_3) = 0; \quad && \mathrm{grad}_3 D_4 = 0
\end{aligned} \tag{11.26}$$

From the non-zero multiplier of σ_2 we get

$$\mathrm{grad}_2\left(\frac{\rho_T}{c} A_4\right) + \frac{1}{c}\frac{\partial}{\partial t}\left[\frac{\rho_V}{c}(A_2 + D_2)\right] \neq 0$$

which, as we will see, is equal to $\mathbf{N}^{\mathrm{FULL}}$.

This set of differential equations, extended with the formula for $\mathbf{M}^D$ yield the four components of $D_{\mu,}$ and the kinetic charge current density ρ_T, which latter depends on the actual relative velocity of the two-charge system. In this case, the electro-kinetic moment (with a mass-dimension divider) will take the form

$$i\gamma_5\left(\frac{\hbar}{c}\boldsymbol{\sigma}, \mathbf{N}^D\right) = i\gamma_5\, \rho_V\left[\frac{\hbar}{c}\boldsymbol{\sigma}, \mathrm{grad}D_4 + \frac{1}{c}\frac{\partial}{\partial t}\mathbf{D} + \frac{\rho_V}{\hbar c}(\mathbf{D}D_4 - D_4\mathbf{D})\right]$$

where the first and third terms in the right side are equal to 0, so the electro-kinetic moment (with a mass-dimension divider) will reduce to

$$i\gamma_5\left(\frac{\hbar}{c}\boldsymbol{\sigma}, \mathbf{N}^D\right) = i\gamma_5\, \rho_V\left(\frac{\hbar}{c}\boldsymbol{\sigma}, \frac{1}{c}\frac{\partial}{\partial t}\mathbf{D}\right)$$

The full electric moment (with an $(m_T - m_V)$ divider) can be written as

$$\gamma_5\left(\frac{\hbar}{c}\boldsymbol{\sigma}, \mathbf{N}^{\mathrm{FULL}}\right) = i\gamma_5\left(\frac{\hbar}{c}\boldsymbol{\sigma}, \mathrm{grad}\rho_T A_4 + \frac{\rho_V}{c}\frac{\partial}{\partial t}(\mathbf{A} + \mathbf{D})\right)$$

whose second component ($i\sigma_2$ multiplied) is real, the first and third components are 0. The electric moment of the interacting particles is directed towards the real component of the spin ($i\sigma_2$). We have not experienced such moment in the classical Dirac theory.[11.12] In contrast to the Pauli–Dirac treatment (Dirac, 1928), this moment is a measurable quantity. Further conclusions are discussed by Darvas (2014).

11.3.9 The Hamiltonian and the Lagrangian of the electromagnetic interaction in the presence of isotopic gravitational and electric charges as well as a kinetic gauge field

We saw that line 5 of Eq. (11.17) yields the Schrödinger wave equation $i\hbar(\partial/\partial t)\psi=\mathbf{H}\psi$, similar to the clue we followed in Section 11.3.2:

$$i\hbar\frac{\partial}{\partial t}\psi = \left[-\rho_T A_4 - \rho_V D_4 - \gamma_5(\boldsymbol{\sigma}, i\hbar c\,\mathrm{grad} - \rho_V \mathbf{A} - \rho_V \mathbf{D}) + \gamma_4 m_V c^2\right]\psi \tag{11.27}$$

and hence the Hamiltonian is

$$\mathbf{H} = -\rho_T A_4 - \rho_V D_4 - \gamma_5(\boldsymbol{\sigma}, i\hbar c\,\mathrm{grad} - \rho_V \mathbf{A} - \rho_V \mathbf{D}) + \gamma_4 m_V c^2$$

The Lagrangian of the interaction field can be constructed from the Hamiltonian. So

$$\mathbf{L} = \rho_T A_4 + \rho_V D_4 - \gamma_5(\boldsymbol{\sigma}, i\hbar c\,\mathrm{grad} - \rho_V \mathbf{A} - \rho_V \mathbf{D}) + \gamma_4 m_V c^2$$

Obviously, this expression differs from the classical one in two terms. The latter include first the three-component **D,** and the fourth component of the vierbein, that is, D_4, on the one hand, and the two isotopic electric charges, on the other hand.

As shown in Section 11.3.2 (and originally in Darvas, 2013b), the condition for obtaining the Schrödinger equation was that $m_T \gg m_V$, that means, an extreme relativistic situation (cf., Section 10). If $m_T \neq m_V$, one can divide the full Eq. (11.17) by $(m_T - m_V)$. One could obtain eq. (11.27) in this way. The division by $(m_T - m_V)$ may cause an increase in the energy of the system when the difference between $m_T - m_V$ approaches to 0 unless the change in **D** does not counterbalance it. This means, first, that **D** must be a monotone function of the velocity; second, we can determine a limit of its monotone increase with the increase of the velocity.

11.3.9.1 The full magnetic and electric moments

We wrote Eq. (11.17) in the form of

$$[W + W^A + W^D - H(m_T - m_V)c]\psi = 0 \tag{11.28}$$

11.12 The fact is that Dirac (1928, p. 619) could not do anything with the electric moment, and so did all but most textbooks following him. The appearance of the kinetic field **D** made possible to calculate the full electric moment.

Close to the rest, division of Eq. (11.17) or (11.28) by $(m_T - m_V)$ makes W and W^A high. This operation gets sense only at far relativistic velocities. Nevertheless, just in the case of low velocities, the role of $H(m_T - m_V)c$ can be neglected, since $(m_T - m_V) \to 0$. What is interesting for us is the role of W^D. Let us divide W^D (11.29)

$$\begin{aligned} W^D &= \hbar\left(\boldsymbol{\sigma}, \operatorname{rot}\left(\frac{\rho_V}{c}\mathbf{D}\right)\right) + \hbar\left(\boldsymbol{\sigma}, \frac{\rho_V^2}{c^2}\left[D_jD_k - D_kD_j\right]\right) + \\ &+ i\hbar\gamma_5\left(\boldsymbol{\sigma}, \operatorname{grad}\left(\frac{\rho_V}{c}D_4\right) + \frac{1}{c}\frac{\partial}{\partial t}\left(\frac{\rho_V}{c}\mathbf{D}\right)\right) + \gamma_5\frac{\rho_V^2}{c^2}(\boldsymbol{\sigma}, D_4\mathbf{D} - \mathbf{D}D_4) \end{aligned} \tag{11.29}$$

by $(m_T - m_V)$:

$$\begin{aligned} &\left(\frac{\hbar}{c}\frac{\boldsymbol{\sigma}}{m_T - m_V}, \rho_V \operatorname{rot}\mathbf{D}\right) + \left(\frac{\hbar}{c}\frac{\boldsymbol{\sigma}}{m_T - m_V}, igC^i_{jk}\frac{\rho_V^2}{c}D_jD_k\right) + \\ &+ i\gamma_5\left[\begin{aligned} &\left(\frac{\hbar}{c}\frac{\boldsymbol{\sigma}}{m_T - m_V}, \rho_V \operatorname{grad} D_4 + \frac{\rho_V}{c}\frac{\partial}{\partial t}\mathbf{D}\right) + \\ &+ \left(\frac{\hbar}{c}\frac{\boldsymbol{\sigma}}{m_T - m_V}, \frac{\rho_V^2}{\hbar c}(\mathbf{D}D_4 - D_4\mathbf{D})\right) \end{aligned}\right] \end{aligned} \tag{11.30}$$

As shown in Section 11.8.3, the last term can be disregarded, since according to our simplifying assumptions D_i and D_4 commute with each other. So one can do with grad D_4 which is 0. We obtain

$$\rho_V\left(\frac{\hbar}{c}\frac{\boldsymbol{\sigma}}{m_T - m_V}, \operatorname{rot}\mathbf{D} + igC^i_{jk}\frac{\rho_V}{c}D_jD_k\right) + i\gamma_5\rho_V\left(\frac{\hbar}{c}\frac{\boldsymbol{\sigma}}{m_T - m_V}, \frac{1}{c}\frac{\partial}{\partial t}\mathbf{D}\right) \tag{11.31}$$

Note that there appear in (11.31) only the potential (Coulomb) charges and the mass difference between the kinetic and potential states. Expression (11.31) can be written also in the form

$$\begin{aligned} &\left(\frac{\hbar}{c}\frac{\boldsymbol{\sigma}}{m_T - m_V}, \mathbf{M}^D\right) + i\gamma_5\left(\frac{\hbar}{c}\frac{\boldsymbol{\sigma}}{m_T - m_V}, \mathbf{N}^D\right) = \\ &= \rho_V\left(\frac{\hbar}{c}\frac{\boldsymbol{\sigma}}{m_T - m_V}, \operatorname{rot}\mathbf{D} + igC^i_{jk}\frac{\rho_V}{c}D_jD_k\right) + i\gamma_5\rho_V\left(\frac{\hbar}{c}\frac{\boldsymbol{\sigma}}{m_T - m_V}, \frac{1}{c}\frac{\partial}{\partial t}\mathbf{D}\right) \end{aligned} \tag{11.32}$$

where $\mathbf{M}^D$ and $\mathbf{N}^D$ are the same, as defined in Eq. (11.25) in Section 11.3.8.3, considering the mentioned omittance. The two terms in the left side of (11.32) are the additional *magneto-kinetic* and the additional *electro-kinetic* moments of the kinetic gauge field of the interaction. As shown at the end of Section 11.3.8.2, in contrast to the classical Dirac theory, in the presence of a kinetic gauge field the electro-kinetic moment cannot be disregarded. Added to the components that are calculated from the electromagnetic field, it may include real components, and in a properly chosen reference frame, it obtained physical meaning. This latter option was not considered in the classical QED (cf., footnote 11.12).

The *full magnetic moment* of the interaction in the presence of the kinetic gauge field was the sum of the magnetic moments that appeared in the QED and that additional one derived earlier:

$$\left(\frac{\hbar}{c}\frac{\boldsymbol{\sigma}}{m_T - m_V}, \mathbf{M}^{\mathrm{FULL}}\right) = \rho_V \left(\frac{\hbar}{c}\frac{\boldsymbol{\sigma}}{m_T - m_V}, \mathrm{rot}(\mathbf{A}+\mathbf{D}) + igC^i_{jk}\frac{\rho_V}{c}D_j D_k\right) \tag{11.33}$$

The *full electric moment*, similarly, was written as

$$i\gamma_5\left(\frac{\hbar}{c}\frac{\boldsymbol{\sigma}}{m_T - m_V}, \mathbf{N}^{\mathrm{FULL}}\right) = i\gamma_5\left(\frac{\hbar}{c}\frac{\boldsymbol{\sigma}}{m_T - m_V}, \mathrm{grad}\rho_T A_4 + \frac{\rho_V}{c}\frac{\partial}{\partial t}(\mathbf{A}+\mathbf{D})\right) \tag{11.34}$$

Note again that in $\mathbf{N}^{\mathrm{FULL}}$, there appears also the kinetic charge density.

These $\mathbf{M}^{\mathrm{FULL}}$ and $\mathbf{N}^{\mathrm{FULL}}$ should commute with the Hamiltonian operator of the interacting two charges (cf., Section 11.3.10.2, and section 5.3 in Darvas, 2013b).

11.3.9.2 The momentum in a kinetic field and the appearance of a virtual "coupling" spin

Assuming a central field, where the vector potential is $\mathbf{A}=0$, but the kinetic field is non-zero, the Hamiltonian of the system of two interacting charges is the following:

$$\mathbf{F} = -\left(p_4 + \frac{\rho_V}{c}D_4\right) + \gamma_5\left(\boldsymbol{\sigma}, \mathbf{p} + \frac{\rho_V}{c}\mathbf{D}\right) + \gamma_4 m_V c$$

This formula differs from the previous one in two terms:

$$-\frac{\rho_V}{c}D_4 + \gamma_5\left(\boldsymbol{\sigma}, \frac{\rho_V}{c}\mathbf{D}\right)$$

Note the following two remarks! On the one hand, these two terms do not give an additional component to the commutation of $\mathbf{F}$ and $(m + (1/2)\hbar\sigma)$: D_1 and D_3 are 0; $\gamma_5\sigma_2 D_2 = -\sigma_2 D_4 = iD_4$, so the two terms annulate (neutralise) each other. On the other hand, while the coefficients of $i\hbar\gamma_5$ in Eqs. (11.26) annulate each other *by rule*, this latter annulation is *accidental* (cf., Pons et al., 2000; Rabinovitz, 2013). This term is real, for $i\sigma_2$ is real. According to Breit (1929, 1932) and further calculations, they give an additional

$$c^2 g\rho_V \frac{i}{4}\hbar\frac{\sigma_2}{m_T - m_V} \tag{11.35}$$

spin to the term $(\mathbf{m} + (1/2)\hbar\boldsymbol{\sigma})$, where g denotes the coupling constant of the electromagnetic interaction, and m_T, m_V are the masses of the interacting electric charges. Since we consider an *interaction* between two electric charges, *this virtual "coupling" spin appears only when there are at least two, interacting electric charges present.* They are present (virtually) only in bound states of two particles. They vanish in all other instances (Breit, 1932). The arrows of the virtual "coupling" spins of two interacting electric charges are directed opposite (anti-parallel), so they compensate each other. Therefore, they are unobservable and do not give any observable additional spin to the two interacting agents. Note also, that this term is becoming large close to the rest, and disappears at high energies. The latter means that the coupling force is much larger between two electric charges near to rest than between those moving with relatively high velocities to each other: at high energies, they become asymptotically free (at least, of the here defined virtual "coupling" spin). The possible existence of such a virtual "coupling" spin was predicted by Darvas and Farkas (2006).

Formula (11.35) shows that the *rest charge*, and the *difference between the kinetic and potential* (rest) *masses* play the determining role in the strength of the interaction between two electrically charged particles.

11.3.10 The field tensors of the EM and the kinetic fields

11.3.10.1 The field tensor of the EM field

According to Darvas (2011), the obtained equations yield the classical QED fields in the absence of a kinetic **D** field. Thus, the elements of the field tensor of the EM field as well as the conserved current are of the same form as we learned in our usual textbooks. This means, our results extend the results accepted in the SM, but do not influence the validity of those equations in the absence of considering a kinetic field. They provide the same conserved quantities as we learned in the semiclassical theory, which means in this instance the electric charge. This conclusion coincides with all said in connection with $J^{(1)}$ in Section 7.4 and Darvas (2011).

11.3.10.2 The field tensor of the kinetic field

The field tensor of the kinetic field (according to Section 7.3) can be obtained as

$$F^{(2)\mu\nu}(x) = \frac{\partial D_{\dot{\rho}}\Lambda^{\rho}_{\mu}}{\partial x_{\nu}} - \frac{\partial D_{\dot{\sigma}}\Lambda^{\sigma}_{\nu}}{\partial x_{\mu}} + D_{\dot{\rho}}\Lambda^{\rho}_{\mu}D_{\dot{\sigma}}\Lambda^{\sigma}_{\nu} - D_{\dot{\sigma}}\Lambda^{\sigma}_{\nu}D_{\dot{\rho}}\Lambda^{\rho}_{\mu}, \tag{11.36}$$

where $\Lambda^{\rho}_{\mu} = \partial_{\mu}\dot{x}^{\rho} = (\partial\dot{x}^{\rho}/\partial x_{\mu}) = \dot{x}^{\rho}_{,\mu}$ (cf., paragraph 1 in Section 7.3).

Similarly, as we obtained the elements of the field tensor for the EM field from the terms in the second line in Eq. (11.17), we can determine the elements of the kinetic field tensor from Eqs. (11.24) and (11.24a). For this reason, we will use the expressions defined for $\mathbf{M}^D$ and $\mathbf{N}^D$ that denote the two components of the field strengths of the field's kinetic potential **D**. From

$$\mathbf{M}^D = \rho_V \mathrm{rot}\mathbf{D} + igC^i_{jk}\frac{\rho_V^2}{c}D_jD_k \quad \text{and}$$

$$\mathbf{N}^D = \rho_V \mathrm{grad}D_4 + \frac{\rho_V}{\hbar c}\frac{\partial}{\partial t}\mathbf{D} + \frac{\rho_V^2}{\hbar c}(\mathbf{D}D_4 - D_4\mathbf{D}) \tag{11.37}$$

one can construct the following tensor:

$$\rho_V F^{\mu\nu} = \begin{bmatrix} 0 & M_3^D & -M_2^D & -i\gamma_5 N_1^D \\ -M_3^D & 0 & M_1^D & -i\gamma_5 N_2^D \\ M_2^D & -M_1^D & 0 & -i\gamma_5 N_3^D \\ i\gamma_5 N_1^D & i\gamma_5 N_2^D & i\gamma_5 N_3^D & 0 \end{bmatrix}$$

where

$$\begin{aligned} \mathbf{M}_i^D &= \partial_j \rho_V D_k - \partial_k \rho_V D_j + igC_{jk}^i \frac{\rho_V^2}{c} D_j D_k = \\ &= \rho_V \left(\partial_j D_k - \partial_k D_j\right) + igC_{jk}^i \frac{\rho_V^2}{c} D_j D_k = \\ &= \rho_V \mathrm{rot}_i \mathbf{D}(\dot{x}) + igC_{jk}^i \frac{\rho_V^2}{c} D_j D_k = \\ &= \rho_V \left[(\partial_{\dot{\rho}} D_k)\lambda_j^{\dot{\rho}} - (\partial_{\dot{\rho}} D_j)\lambda_k^{\dot{\rho}}\right] + igC_{jk}^i \frac{\rho_V^2}{c} D_j D_k \end{aligned} \tag{11.38}$$

and

$$\begin{aligned} \mathbf{N}_i^D &= \partial_i \rho_V D_4 + \frac{1}{\hbar c} \partial_t \rho_V D_i + \frac{\rho_V^2}{\hbar c} D_i D_4 - \frac{\rho_V^2}{\hbar c} D_4 D_i = \\ &= \rho_V \left(\partial_i D_4 + \frac{1}{\hbar c} \partial_t D_i\right) + \frac{\rho_V^2}{\hbar c} (D_i D_4 - D_4 D_i) \end{aligned} \tag{11.39}$$

Considering that grad D_4=0 and D_4 commutes with D_i:

$$\begin{aligned} \mathbf{M}_i^D &= \rho_V (\mathrm{rot}_i \mathbf{D}(\dot{x}) + igC_{jk}^i \frac{\rho_V}{c} D_j D_k) = \\ &= \rho_V \left[(\partial_{\dot{\rho}} D_k)\lambda_j^{\dot{\rho}} - (\partial_{\dot{\rho}} D_j)\lambda_k^{\dot{\rho}}\right] + igC_{jk}^i \frac{\rho_V^2}{c} D_j D_k \end{aligned} \tag{11.40}$$

and

$$\mathbf{N}_i^D = \frac{\rho_V}{\hbar c} \partial_t D_i \tag{11.41}$$

11.3.10.3 The curvature of the connection field

The curvature of the connection field can be read from the coefficient of the covariant extension of the matrix terms in $F^{\mu\nu}$ (11.37)–(11.39). $\mathbf{M}^D{}_i$ can be written also in the form (11.40), where the last two terms define a covariant commutation of the elements D_i:

$$\mathbf{M}_i^D = \rho_V (\mathrm{rot}_i \mathbf{D}(\dot{x}) + ig\Gamma_{jk}^i D_j D_k).$$

Here $\Gamma^i_{jk} = C^i_{jk}(\rho_V/c)$ denotes that Γ depends only on constants, while D_i depend on the $\dot{x}^\mu$ four-velocity components. The latter $\dot{x}^\mu(x_\nu)$corresponds to the functions marked by Dirac (1962) as y^Λ_ν through which he defined the metric of the field.[11.13,11.14] The metric of the field is much simpler than we expected, while the velocity dependence is transferred to the components of the **D** field.

11.3.11 The Lorentz force in the presence of a kinetic field

Similar to the classical EM model (cf., Section 11.3.1), in the presence of isotopic electric charges and a kinetic field, the Lorentz force – that we discussed in its classical form in Section 3.5 – can be written in the following form now:

$$F^\mu = F^{\mu\nu}\frac{1}{c}j_\nu = \frac{1}{\rho_V}\begin{bmatrix} 0 & M_3^D & -M_2^D & -i\gamma_5 N_1^D \\ -M_3^D & 0 & M_1^D & -i\gamma_5 N_2^D \\ M_2^D & -M_1^D & 0 & -i\gamma_5 N_3^D \\ i\gamma_5 N_1^D & i\gamma_5 N_2^D & i\gamma_5 N_3^D & 0 \end{bmatrix}\begin{bmatrix} \rho_T\frac{\dot{x}1}{c} \\ \rho_T\frac{\dot{x}2}{c} \\ \rho_T\frac{\dot{x}3}{c} \\ i\rho_V \end{bmatrix} =$$

$$= \frac{1}{\rho_V}\begin{bmatrix} M_3^D\rho_T\frac{\dot{x}2}{c} - M_2^D\rho_T\frac{\dot{x}3}{c} + \gamma_5 N_1^D\rho_V \\ -M_3^D\rho_T\frac{\dot{x}1}{c} + M_1^D\rho_T\frac{\dot{x}3}{c} + \gamma_5 N_2^D\rho_V \\ M_2^D\rho_T\frac{\dot{x}1}{c} - M_1^D\rho_T\frac{\dot{x}2}{c} + \gamma_5 N_3^D\rho_V \\ i\gamma_5 N_1^D\rho_T\frac{\dot{x}1}{c} + i\gamma_5 N_2^D\rho_T\frac{\dot{x}2}{c} + i\gamma_5 N_3^D\rho_T\frac{\dot{x}3}{c} \end{bmatrix} =$$

$$= \frac{1}{c\rho_V}\begin{bmatrix} M_3^D\dot{x}^2 - M_2^D\dot{x}^3 & c\gamma_5 N_1^D \\ -M_3^D\dot{x}^1 + M_1^D\dot{x}^3 & c\gamma_5 N_2^D \\ M_2^D\dot{x}^1 - M_1^D\dot{x}^2 & c\gamma_5 N_3^D \\ i\gamma_5 N_1^D\dot{x}^1 + i\gamma_5 N_2^D\dot{x}^2 + i\gamma_5 N_3^D\dot{x}^3 & 0 \end{bmatrix}\begin{bmatrix} \rho_T \\ \rho_V \end{bmatrix} = \frac{1}{c\rho_V}H^{D\kappa l}\rho_l$$

where (κ=1,. . ., 4) (l=1, 2), or in the form

11.13 While the Dirac equation – introduced and discussed first in his 1928 and 1929 papers – is presented in almost all usual textbooks on QED and field theory, his extension published in 1962 is mentioned rarely (cf.,Weinberg, 1997).

11.14 Fabbri (2013) studied also the metric of Dirac's field theory with kinematic conditions, in a similar, but also a little bit different context.

$$F^{\mu} = \frac{1}{c}\begin{bmatrix} M_3^D\dot{x}^2 - M_2^D\dot{x}^3 & c\gamma_5 N_1^D \\ -M_3^D\dot{x}^1 + M_1^D\dot{x}^3 & c\gamma_5 N_2^D \\ M_2^D\dot{x}^1 - M_1^D\dot{x}^2 & c\gamma_5 N_3^D \\ i\gamma_5 N_1^D\dot{x}^1 + i\gamma_5 N_2^D\dot{x}^2 + i\gamma_5 N_3^D\dot{x}^3 & 0 \end{bmatrix}\begin{bmatrix} \frac{\rho_T}{\rho_V} \\ 1 \end{bmatrix} \tag{11.42}$$

It is obvious in the latter form that the isotopic electric charges do not concern the electric moment. Their ratio plays the role of a coefficient to the magneto-kinetic moment only. This ratio depends only on the Lorentz transformation, in which there appears the relative velocity of the two interacting charges to each other. This expression for the Lorentz force shows that our Γ curvature obtained for the kinetic field is in its form similar to the Γ curvature for the EM field as determined by Landau and Lifshitz (1967, §85).

The kinetic addition to the Lorentz force can be defined with the use of the $F^{\mu\nu}$ applied in line 2 of the 1st equation in this subsection (11.3.11):

$$F^{\mu} = \frac{1}{c\rho_V}\begin{bmatrix} \left[M_k^D \times \dot{x}^j\right] & c\gamma_5 N_i^D \\ i\gamma_5 N_i^D \dot{x}^i & 0 \end{bmatrix}\begin{bmatrix} \rho_T \\ \rho_V \end{bmatrix} = \frac{1}{c\rho_V} H^{D\kappa l}\rho_l \tag{11.43}$$

This expression for the Lorentz force indicates that the weak intermediate bosons can be derived from the [4×2] matrix in the first square bracket [ ]: the photon γ, with mass zero, is associated with H^{D22}, $W^{\pm}$ with H^{D12} and H^{D21}, while Z^{o} with H^{D11}. Please note the asymmetry between H^{D12} and H^{D21}, which confirms the assumption by C. Møller (1931), and what was indicated by S. Weinberg (1967) in another way. Note also that we indicated the unification of the *electromagnetic* and the *weak interactions* in a different way than Weinberg did.

11.3.12 The conserved currents and the conserved isotopic electric charge spin

In the possession of the Lagrangian and the field tensor, with the help of the D_i kinetic potential field components, one can apply the Eq. (11.11) obtained in Section 11.3.2:

$$J_{\alpha}^{(2)\nu}(x) = ig\left[\frac{\partial L}{\partial(\partial_{\mu}A_{4k})}(T_{\alpha})_{kl}A_{4l}(\dot{x})\Lambda_{\mu}^{\nu} - C_{\alpha\beta}^{\gamma}D_{\dot{\omega}\beta}(\dot{x})\Lambda_{\mu}^{\omega} \times F_{\gamma}^{(2)\mu\nu}(x)\right] \tag{11.44}$$

In the instance (of the electromagnetic interaction) α=1, $(T_{\alpha})_{kl}$ is a unitary matrix, and $C^{\gamma}{}_{\alpha\beta}$ are the three structure constants for the kinetic field that commute the generators of the $SU(2)$ symmetry group. $J^{(2)\nu}$ are the components of the conserved isotopic electric charge spin current, which include the contribution of the kinetic **D** field (cf., the second term on the right side)[11.15] ($\mathbf{J}^{(2)}=\mathbf{J}^{D}$). We have introduced the **D** field – which is shown to be responsible for the IFCS transformation – to counteract the dependence of a V transformation on $J^{D_{1,2,3}}$ (Section 8.2). The field equations, which are satisfied by the 12 independent components of the **D** field, and their interaction with any field that carries IFCS, are unambiguously determined by the defined current and the covariant $F^{(2)\mu\nu}$-s constructed from the components of **D**. Considering a general Lorentz- and gauge-invariant Lagrangian, we obtain from the equations of motion that $J^{D_{1,2,3}}$ and J^{D_4} are, respectively, the IFCS current density and isotopic electric charge spin (Δ) density of the system. The total isotopic electric charge spin is as shown in Section 8.2:

$$\Delta = \frac{i}{g}\int J^{D_4}\, d^3x \tag{11.45}$$

is independent of time and independent of Lorentz transformation. $J^{D\mu}$ does not transform as a vector, while Δ transforms like a vector under *rotations in the isotopic electric charge spin field.* The Δ_{el} quanta of the **D** field (for EM) can be determined by insertion in the expression J^{D_4}:

$$\Delta_{el} = -\int\left[\frac{\partial L}{\partial(\partial_\mu A_4)}A_4\partial_\mu\dot{x}^4 - \gamma_5 C^\gamma D_{\dot{\rho}\gamma}\partial_i\dot{x}^\rho \times N_i\right]d^3x \tag{11.46}$$

$$(C^\gamma = 0, \pm 1;\quad \gamma = 1, 2, 3,\quad \mu = 1, \ldots, 4)$$

We showed (Section 8) that beside the conserved electromagnetic current (which provides the conservation of the electric charge, and the γ quantum of the EM field), there exists another conserved current, J^{D_4}, in the presence of a kinetic gauge field **D** (see also Section 11.3.13).

11.3.13 Quantisation

The coupling of a conserved quantity in a space–time dependent field (which coincides with one of our known SM physical fields) with another (in a kinetic, that is, velocity dependent (Akhmedov and Smirnov 2011)) **D** gauge field indicates that the derived conservation verified the invariance between two isotopic states of the field-charges, namely between the potential and the kinetic electric charges. Remember that the conserved isotopic electric charges belong to the electromagnetic field, while Δ represents a single quantity belonging to the kinetic gauge field **D.**

11.15 Similar attempts (like our in the velocity space) were made by Pons et al. (2000) in the phase space (with a particular mapping from the configuration space to phase space), and they anticipated the quantization of the models.

In short, in the presence of kinetic fields, we have two conserved currents that are effective simultaneously (Section 8.3). The kinetic gauge field **D** is present simultaneously with the interacting matter (A_4) and gauge (**A**) fields.[11.16] The presence of **D** corresponds to the property of the electric field-charges, since they split into two isotopic states. And (analogously to the isotopic spin) these two states are given the name *isotopic electric charge spin* what we denoted by Δ_{el}. The source of the isotopic electric charge spin (Δ_{el}) is the field $D_4(\dot{x})$, in interaction with the kinetic gauge field **D**.

The physical meaning of Δ_{el} can be understood by the specification of the transformation group associated with the **D** field, which describes the transformations of the isotopic electric charges. They can take two (potential and kinetic) isotopic density states ρ_V and ρ_T in a simple unitary abstract space. Their symmetry group is isomorphic with $SU(2)$, which can be represented by 2×2 τ_α matrices ((Mills, 1989) or (Section 5.2)). Apologies for some repetition (cf., Section 8.4), there are three independent τ_α that may transform into each other, following the rule $[\tau_\alpha, \tau_\beta] = iC^\gamma_{\alpha\beta}\tau_\gamma$, where the structure constants can take the values 0, ±1. Let τ_1 and τ_2 be those which do not commute with τ_3;. they generate transformations that mix the different values of τ_3,while this "third" component's eigenvalues represent the members of a Δ_{el} doublet. For the IFC compose a charge density doublet of ρ_V and ρ_T, the field's wave function can be written as

$$\psi = \begin{pmatrix} \psi_T \\ \psi_V \end{pmatrix} \tag{11.47}$$

Expression (11.47) is the wave function for a single electron which may be in the "potential state", with amplitude ψ_V, or in the "kinetic state", with amplitude ψ_T. Note that the potential (Coulomb) charges behave like *corpuscles*, while the kinetic (Lorentz type) charges like *waves*. This complementary double behaviour (formulated first by Bohr in 1927, then discussed by Bohr (1937) became a subject of studies again (cf., Rabinovitz, 2013)). ψ in (11.47) represents a mixture of the potential and kinetic (corpuscle and wave) states of the electric charge, and there are τ_α that govern the mixing of the components ψ_V and ψ_T in the transformation. τ_α (α=1, 2, 3) are representations of operators which can be taken as the three components of the isotopic electric charge spin, namely Δ_1, Δ_2, Δ_3 that follow the same (non-Abelian) commutation rules as do the τ_α matrices, $[\Delta_1, \Delta_2] = i\Delta_3$, and so on. These operators represent the charges of the isotopic electric charge spin space, and ψ are the fields on which the operators of the gauge fields act.

The quanta of the **D** field should carry isotopic electric charge spin Δ_{el}. The Δ_{el} doublet, as a conserved quantity, is related to the two isotopic states of the electric

11.16 Concerning the parallel presence of a scalar and a kinetic gauge field, Sarkar et al. (2013) enlarged the configuration space by including a scalar field additionally, and taking anisotropic models into account too. They also investigated whether Noether's symmetry holds in the known form alone or does not, which means whether the single Noether current should be extended (let us add: with another current).

charges, and the associated operators (Δ_i) induce transitions from one member of the doublet to the other.

The *invariance between* ρ_V *and* ρ_T (what is ensured by the conservation of Δ_{el}), and their ability to swap, *means* also that *they can restore the symmetry in the physical equations which appeared to get lost when we replaced the general* ρ *by their isotopes* ρ_V *and* ρ_T.

These considerations substantiated the prediction of quanta (dion δ_{el}) in the **D** field associated with the electromagnetic field.

Summarising the above results, in short, when we introduced the distinguished two kinds of isotopic electric charges in our physical equations we reached a limit: we obtained equations where certain symmetries of the traditional equations were distorted. That was not in accordance with our experience. Then, we derived a conservation law for the newly introduced quantity, the isotopic electric charge spin (Δ_{el}), and its invariance (applied together with the Lorentz invariance) restored the lost symmetry (cf., Sections 11.3.5–11.3.6). The isotopic electric charge spin should exist in an above-defined gauge field **D**.

We predicted in Section 11.3.2 that such a gauge field must have quanta that carry the isotopic electric charge spin Δ_{el} (cf., footnote 9.3). Exchange of these quanta should mediate between the interacting field-charges, so that switch the emitter electric charge $\rho_V \rightarrow \rho_T$, and the recipient electric charge from $\rho_T \rightarrow \rho_V$ and *vice versa*. This holds because according to field theories, any conserved property resumes the existence of a mediating boson. In Section 9.3, we demonstrated that in this case too, a mediating boson – called "dion" – must belong to each IFC pair. The conservation of Δ_{el} involves the prediction of a dion associated with the electromagnetic field, a boson brother of the photon. The HySy theory describes the necessity for the existence of these dions.

We denoted the *predicted* quanta of the **D** field by δ (Section 9.3). The δ_{el} quanta (dions) carry the Δ_{el} (isotopic electric charge spin as a physical property: the charge of the **D** field). Since all the quantum numbers of the interacting field-charges are mediated by the respective SM mediating bosons, the quantum numbers of the dion (δ) are 0 but the Δ (and, of course, they carry mass according to Section 9.3). We can agree in a free convention for the sign of Δ_{el} to be + or – (½) according to whether it switches the IFCS from a potential state to a kinetic or back.

11.3.14 Observation of a dion (a δ boson)

The observation of dions is possible in high energy collisions. The effect of the kinetic field is more apparent at high energies (high velocities). At higher energies, the mass exchange between the two interacting IFC ($m_T - m_V$) is higher, due to the increasing relativistic difference between m_T and m_V. The effective cross section increases with the increase of the energy of the interaction, and the probability to find the searched dions increases with the energy. However, at too high velocities (in case of p–p and

e–e collisions over cca. 0.998 c) the mass of the respective dion and m_T become so close to each other that they will be indistinguishable, at least by their masses.

Observation of the quantum δ_{gr} for the gravitational field seems unplausible since one has never seen any graviton. Observation of the electric δ_{el} and the three weak dions is more probable, but the probability of the realisation is low, although not excluded.

According to its interpretation (cf., Section 10.3), the mass of a dion (δ) is the difference between the boosted (dressed, m_T) and the rest (invariant, bare, m_V) masses. The mass of this boson is independent of the type of interaction (gravitational, electromagnetic, weak, strong): $m(\delta) = m_T - m_V = (\kappa - 1)m_V$ where $\kappa = 1/\sqrt{1 - v^2/c^2}$.

Remember that the mass of a dion has no fixed value. Dions differ from any other known (SM) boson in the velocity dependence of their mass. Their mass approaches 0 at low velocities, and it approaches the dressed mass of the particle near the speed of light. Therefore, one cannot observe a dion's mass either at low energies (low colliding velocities, cf., Figures 10.3 and 10.4) and cannot separate its mass from the dressed, m_T mass of the colliding particle at very high energies (high velocities). For example, in the case of a p–p collision over cca. 100 GeV, the mass of the respective dion nearly coincides with the dressed mass of the colliding proton. The approximate indistinguishability of the proton's and the dion's masses at very high energies indicate to measure dion's masses in p–p collisions at the range between 100 MeV and 10 GeV energies. One can detect and measure dions' masses in e–e collisions even at much lower energies, below 100 MeV.

The process in a p–p collision is as follows:

$$p_1(v) \to p_1(0) + \gamma + \delta; \quad p_2(0) + \gamma + \delta \to p_2(v)$$

In words, an accelerated proton $p_1(v)$ with velocity v collides another p_2 in rest (both the velocity of the accelerated p_1 and the 0 velocity of the p_2 in rest are measured in laboratory system). Their masses before the collision are $m_T{}^{(p1)}$ and $m_V{}^{(p2)}$, respectively. $m_T{}^{(p1)} = \kappa m_V{}^{(p1)}$ and $m_V{}^{(p2)} = m_0{}^{(p2)}$. The situation is similar in the case of e–e collision.

During the interaction, the approaching p_1 emits a photon and a dion, and the p_2 in rest absorbs them. Assuming elastic scattering, the approaching p_1 loses its velocity and its dressed mass, which, conveyed by the dion, is given to the target p_2 that accelerates to the former velocity v of the partner.

A dion with the mass $m(\delta) = m_T - m_V = (\kappa - 1)m_V$ should be detected as a short life resonance at the moment of the collision. The expected masses of the dions, depending on the relative velocity of the colliding particles, calculated at a few velocities are as follows:

v/c	0.943	0.98	0.99	0.995	0.998	0.999
$m(\delta)/m_0$	2	4.025	6.089	9.013	14.82	21.37

e–e collisions could be studied in the 0.9 *c*–0.99 *c* velocity range already at 10–100 MeV energies. *p–p* collisions could be studied at higher energies, up to 10 GeV.

It seems more probable to demonstrate the existence of the eight strong dions, indirectly, but that will be the theme of another section (cf., Section 11.5).

11.3.15 Concluding remarks to Section 11.3

(1) The potential mass and charge appear in the Coulomb potential and do not change with increasing velocity. The kinetic mass and charge appear in the kinetic part of the Hamiltonian, and change with the velocity. (Other extensional terms in the Hamiltonian behave similarly.) Therefore, the two kinds of constituents of the Hamiltonian cannot be contracted, like in the classical Dirac QED. The difference between the potential and kinetic charges and masses were considered by Weinberg (1967) integrated into the two kinds of coupling constants (cf., his eq. (15), p. 1265).

Weinberg (1967) described a renormalisable model of spontaneously broken symmetry between electromagnetic and weak interactions, by a two-step perturbation, introducing the photon and the weak intermediate bosons as gauge fields. At the end of his paper, he put the question: " . . . what happens if we extend it to include the couplings of A_μ and B_μ to the hadrons?" (p. 1266). That was a first step towards the unification of his combined electroweak theory with the strong interaction. Our discussion in this section derived the electromagnetic theory by the help of another two-step perturbation (A_μ and D_μ), starting from a kinetic extension of the Dirac equation.

(2) The novelties in the discussion in this section can be most easily understood by starting with Eq. (11.28):

$$[W + W^A + W^D - H(m_T - m_V)c]\psi = 0$$

The last term has the form of a Schrödinger equation (cf., Eq. (11.27) which plays an independent role with conditions when the first three terms can be neglected (cf., (5b) later). This has appeared in the Dirac equation never before!

(3) The third term (W^D) is a contribution to the electric moment. (According to the author, Dirac (1928) was mistaken, when he wrote that the electric moment was imaginary. In fact, it was only one component of σ which was imaginary. But the importance lies not in this.) This additional term (W^D) to the electric moment allowed us to determine the real part of the electric moment. Making the imaginary parts equal to 0, we got a set of differential equations (11.27) to explicitly identify the form of the unknown quantities (Section 11.3.8.3.2). So, **D** is fully determined.

(4) Dirac was aware that his theory was an approximation. Many of his followers were not. He tried to extend his theory twice (1951 and 1962). He failed both times to reach a final solution.

C. Møller (1931) showed the asymmetry between the roles of the interacting isotopic electric charges treated in this book (cf., Section 3). Unfortunately, Bethe (and Fermi, 1932) symmetrised it mistakenly.

This asymmetry was used also by S. Weinberg (1967) in his chiral electroweak unification theory without any reference to C. Møller.

(5a) To justify the approximative character of the "semi-classical" Dirac theory, let us first consider an unperturbed, classical case when m_V=m_T. In this case, the last three lines in Eq. (11.17) vanish, and we get back to the "semi-classical" Dirac equation. All experimental data were obtained for this slightly relativistic situation. Now, we learned that was only an approximation when one disregarded the additional terms. However, this extended Dirac equation includes, as a semi-classical boundary case, the original equation (cf., line 1 in Eq. (11.17)).

(5b) In an "extremely" relativistic case, when $m_T \gg m_V$, the first four lines of Eq. (11.17) can be abandoned. In this case, the situation reduces to the fifth line, which is identical to the Schrödinger equation. Its validity has also been confirmed many times.

(5c) Now, the situation is similar to that in the time of the first formulation of Planck's quantum hypothesis for the black body radiation, set up in the Fall of 1900. We had experimental tests for the two ends of a kinetic energy scale. But we could not have to fit them in the middle. Now we can!

Therefore, the most interesting situation is, when $m_T > m_V$, but not "too much". This situation is discussed in Sections 11.3.8–11.3.8.3.2 in detail. Lines 3–4 (W^D) in Eq. (11.17) get importance in this instance. This is the part of the equation that was disregarded in Dirac's (1928) paper (and in the majority of textbooks, which refer to it, up to now), and this is to which he refers so that his theory is "only" an approximation (cf., remarks at the end of Section 11.3.2). This situation has not been tested, because it has not been discussed. This test should be executed.

(6) Based on the calculation of the components of the **D** field in Section 11.3.8.3, we have determined the extended, full magnetic and kinetic moments in an electromagnetic interaction. We can calculate a new Hamiltonian and Lagrangian of the interaction (cf., Section 11.3.9 and Darvas, 2014), and the full magneto-kinetic and electro-kinetic moments (cf., Section 11.3.9.1 and Darvas, 2014) on that basis. Both are real measurable physical quantities. This restored the physical meaning of the electric moment rejected by Dirac in (1928) (cf., footnote 11.12).

(7) The determination of the momenta of the kinetic field and application of the commutation rules provided an additional "virtual" coupling spin (Eq. (11.35)), which is unobservable, because it occurs only in the interaction between two charges, never in single particles, and is anti-parallel at the

two interacting agents' sides, that means, the two values compensate each other (Section 11.3.9.2).

(8) The appearance of this additional, "virtual" coupling spin represents an accidental symmetry (discussed in Weinberg (2012, pp. 13–14; and 2011)). The existence of such an additional coupling spin was predicted in Darvas, Farkas (2008), but its exact value is derived in Section 11.3.9.2.

(9) There are the *rest charge density* and the *difference between the kinetic and potential* (rest) *masses* that play determining role in the strength of the interaction between two electrically charged particles.

(10) The value of the isotopic electric charge spin is determined in Section 11.3.12.

(11) The invariance between the potential and kinetic charges (what is ensured by the conservation of the isotopic electric charge spin) and their ability to swap means also that they can restore the symmetry in the physical equations that appeared to get lost when we replaced the general charge by their two isotopic states (Section 11.3.13).

11.4 Mechanism of the Δ exchange in weak interactions

The weak interaction (coupled with the electromagnetic) is part of the united electroweak interaction. In this instance, the electric charge appears as one of a four-member multiplet. Moreover, it is extended with two other families, where the muon and tau leptons play the same role as the electric charge. The other three agents of the mentioned multiplet that are subject to weak interaction are the flavours of the neutrinos, lower and upper quarks, in all the three families. The weak interaction among the listed agents is mediated by the three weak current bosons (W^+, W^-, Z). Together with the photon, the four electroweak mediating bosons are the consequence of the $SU(2)\otimes U(1)$symmetry. The combined symmetry of the electroweak interaction means that the agents form multiplets individually by families (although as members of the combined group). Thus the electron, muon and tau form doublets with their respective neutrinos, and hadrons form multiplets with a quantum mixture of their other flavoured brothers.

In the framework of the IFCS model, all members of these multiplets have got a twin pair, so that if they were considered in the potential state (sources of their respective field) in the SM, their twin sibling has a kinetic state IFC, and the same holds for the opposite configuration. Switching between the twin siblings, which means between the two opposite Δ states of the individual flavours during the interaction, is mediated by a corresponding δ_W (weak dion).

The weak current bosons and δ_W act (anti)parallel to Z^o, W^+, and W^-, as in the above-discussed two types of interactions (Figures 11.7a-c).

For the sake of illustration, the interactions mediated by Z and by W^+ are presented on the example of nucleons (p, n), although the agents that interact are their

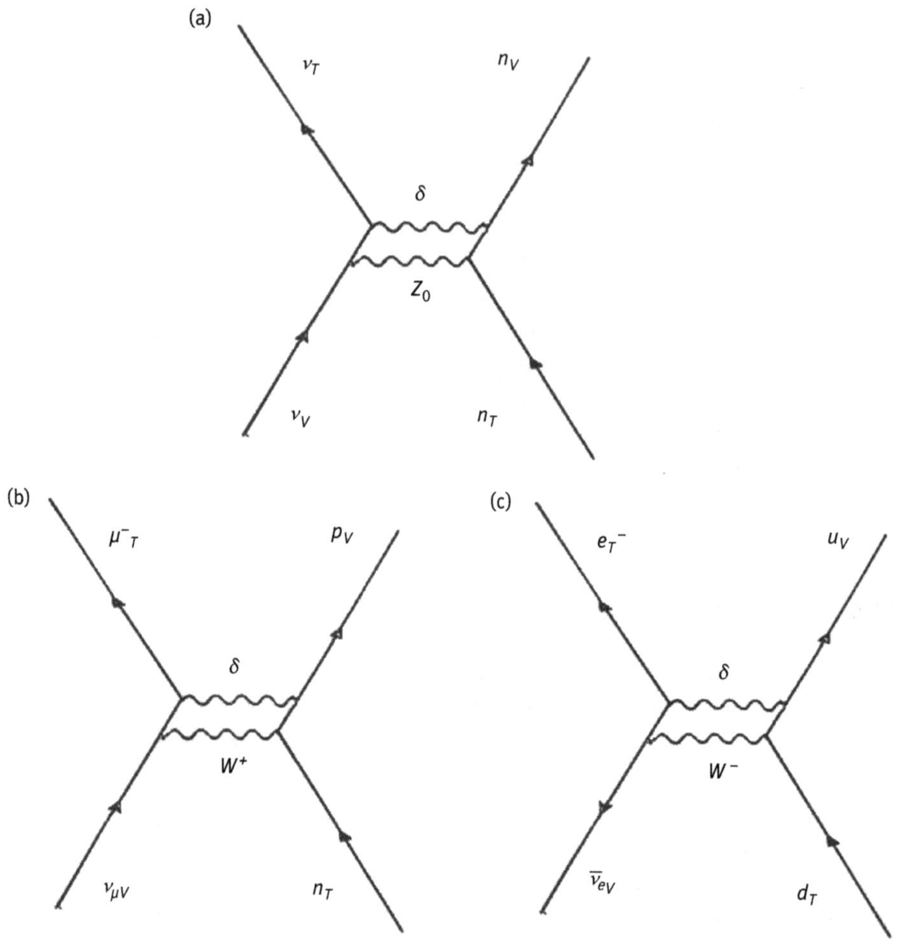

Figure 11.7a–c: Two boson exchange in interaction between the field-charges of weak fields. (δ denotes the dion, Z_0, W^+ and W^- denote the weak current bosons.)

quark components (*u*, *d*). In the example of the W^- mediated interaction, the agents are – in accordance with the reality – quarks (*u* and *d*) themselves. Note that the agents of weak interaction possess not only weak charges (cf., Section 9.2), and are subjects not only of weak interaction; the exceptions are the neutrinos that participate only in electroweak interactions. The role of the electric charge in electromagnetic interaction was discussed in Section 11.3. Flavoured hadrons are also coloured and accordingly may interact strongly, as we will show in Section 11.5.

11.5 Mechanism of the *Δ* exchange in strong interactions

Field charge of the strong interaction is colour. We assume that colour appears also in both potential and kinetic states (cf., Section 9.2). However, due to their confinement, at least in everyday experimental circumstances, we cannot observe them in kinetic states, what supposes free motion. And yet, on the one side, we can imagine that coloured particles may oscillate within their cage limited by confinement; on the other side, they can be freed at very high energies. Since they cannot be assumed as static objects, they must have a kinetic state too. Therefore, similar to the mass, electric charges, weak charges, the colour should also have Δ, and appear both in potential and kinetic states ⌝. The exchange of δ_C bosons (colour dions) should take place in a similar way like the exchange of gluons is interpreted, parallel (or antiparallel) with those. In the presence of three interacting particles, there has got special emphasize to switch to an opposite IFC state (by the exchange of a G_0 gluon and a δ_C dion) before interacting with another colour. (See illustration on the examples of the π^+ and the proton in Figure 11.8a–b.)

In the example of a π^+ meson, the situation can easily read from the left diagram.

The gluon exchange takes place like in the SM model with the difference that (anti)parallel with a SM gluon a δ_C is also exchanged. In the example of a proton (time flow is arrowed up), the red potential and the blue kinetic states of two u quarks exchange a $G_{r\bar{b}}$ gluon and a δ_C colour dion. At the next step, a blue potential u quark and a green kinetic d quark exchange a $G_{r\bar{g}}$ gluon and a δ_C colour dion. Following this step, the two u quarks get both in a kinetic ⌝ state (green and red), so according to the HySy, it is forbidden for them to change a dion, as it would be expected in the SM where they did not owe an IFCS (Δ) charge. So, before the next gluon exchange between the two u quarks, first, the u and the d quarks must commute between their states. In this order, they should exchange a G_0 gluon and a dion. The exchange of G_0 will not change their colour, but the dion exchange commutes their ⌝ states.

Now, the g_V colour$_{\text{state}}$ u quark can exchange a $G_{g\bar{b}}$ gluon and a dion with the b_T colour$_{\text{state}}$ u quark. In the next step in the SM, the blue u quark would be expected to change a gluon with the red d quark. Since, in our instance, they are both in a kinetic state, this is forbidden in the HySy. First, the blue u quark must change its kinetic state to potential state. In this order, it should exchange a G_0 gluon and a dion with the other u quark, which is in a potential state. Again, the exchange of G_0 will not change their colour, but the dion exchange commutes their ⌝ states. Now, the left illustrated u quark gets in a green kinetic (g_T) state, and the middle illustrated u quark in a blue potential (b_V) state, so the latter can exchange a gluon and a dion with the d quark. The b_V state u quark changes a $G_{r\bar{b}}$ gluon and a δ_C dion with the red kinetic state r_T d quark. The permanent gluon and dion exchange continues so on. The observer may record that the number of gluons produced in the process nearly doubles when compared to the number expected in the SM.

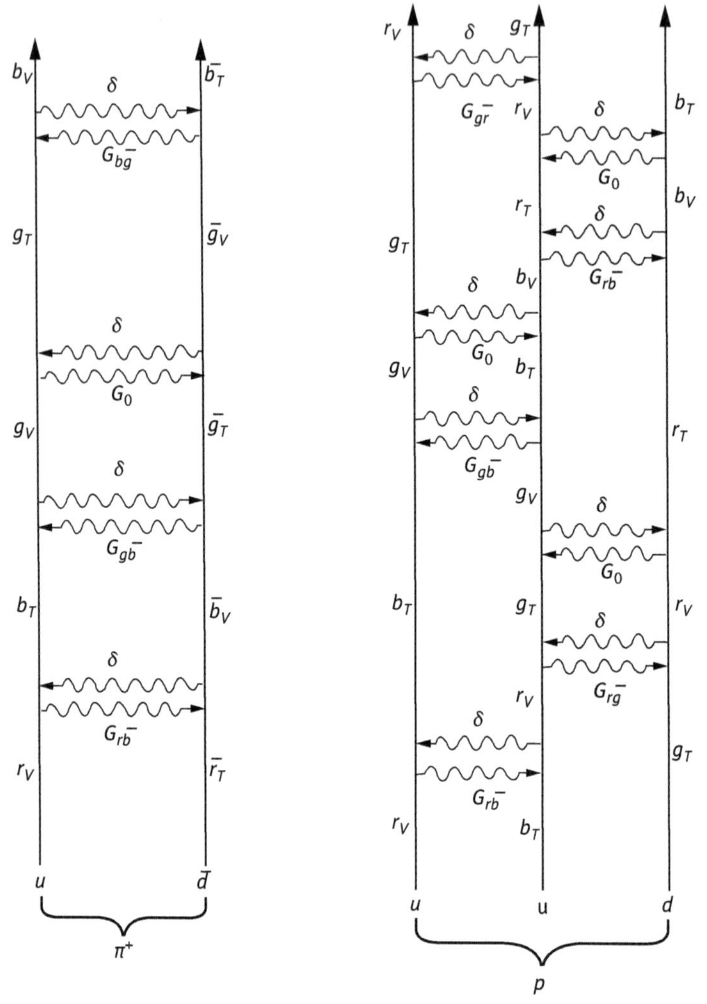

Figure 11.8a–b: Two boson exchange in strong interaction between colours, on the example of the π⁺ mezon (left) and the proton (right).

12 SUMMARY

12.1 The birth and childhood of IFC hypersymmetry

The theory of *isotopic field-charges* (*IFC*) was elaborated by the author of this book, at the Symmetrion. The idea conceived in his mind on January 2001. (It was no more than a tempting thought at that time, it got its name later.)

Dual forms of mass, as the charge of the gravitational field, were present in physics actually since Newton, however, one did not consider to introduce the distinction between the gravitational and inertial masses in the equations of physics. There would have no much sense to make this near to rest. The interpretation of the "principle of *equivalence*" as *identity* (that was suggested also by Einstein one time) strengthened the feeling of superfluity to distinguish them in many scholars. The reason to speak at least about their equivalence was just their difference since only two qualitatively different entities can be equivalent in their quantities. The essential difference between them is in their behaviour under a velocity boost. The history of their distinction, beginning with the papers by G. Mies in 1912, is discussed in a paper read at PIRT-2017 (Darvas, 2017c) and in Section 2.

Concerning the electric charge as a source of the electromagnetic field, we knew since the nineteenth century that the Coulomb charge and the Lorentz charge in the four-vector current were not fully identical quantities, and yet, one did not deal with their distinction. C. Møller (1931) showed already 89 years ago that an asymmetry between the roles of the two interacting agents demonstrates itself in the scattering matrix of two charges, but Bethe (in collaboration with Fermi) (Bethe, Fermi, 1932) artificially "symmetrised" (sic!) the asymmetric matrix, and that high-handed measure was not questioned by the physicists' community until the early 2000s, due to the later prestige of the two authors.

These disregarded qualitative differences between the sources of the individual interaction fields were interpreted as *isotopic states of the field-charges of the given fields* (cf., Section 4). We will explain the causes of the name later.

At that time (the early 2000s), the standard model (SM) became completed (except for the demonstration of the Higgs boson); one could see the limits of its applicability, and the idea of the necessity of a new physics (NP) matured, expected at least in a future that appeared to be far ahead. By that, most physicists assumed to draw the outlines of the NP based on string theories for long, and the conjecture of supersymmetry (SUSY) had got the most attention for about a decade. Alternatives were either thrown into the shade or, were/are sought, similar to the SUSY, on string theoretical bases.

As the idea of IFC ripened to concepts, so presented itself the possibility to use it to an NP in the first years of the 2000s, although one could not foresee then that IFC and SUSY exclude each other.

https://doi.org/10.1515/9783110713183-012

The first consultations were performed by the author with Yuval Ne'eman at the Wigner centenary conference in Pécs 2002, then, following written correspondence, continued at the Symmetry Festival 2003 held in Budapest (in the company of Larry Gould) again. Ne'eman supported to develop the idea.

Having nursed the idea for 3 years, it was broached that if we accept the existence of IFCs, why could not they interact with each other. Moreover, while the gravitational mass and the Coulomb charge appear in the potential energy component of the Hamiltonian of an interacting system, the inertial mass and the Lorentz charges appear in its kinetic component, this would admit a physical mechanism, in which, widely unusually, the components of a Hamiltonian may interact with each other. Even if this happens, it involves further physical consequences. Furthermore, as far as one of the members of the IFC twins appears only in the kinetic energy, then this isotopic member of the field-charges should be in rapport with a velocity-dependent field.

Ne'eman proposed to formulate the mathematical description of the possible mechanism first and then publish its physics. Of course, the two could not be fully separated. Ne'eman considered first of all the algebra (group) that describes the transformation of the field-charges into each other, which are considered isotopes of each other. Unfortunately, one succeeded to describe this group only much later, in the summer of 2014 (Darvas, 2015b, c). However, it succeeded to prove the existence of the possible mechanism analytically already in the autumn of 2004. Even it was surprising first that the proof led to a new conserved property that was named, based on analogies, *isotopic field-charge spin* (*IFCS*).

In short, one started off that if IFC sibling pairs can be distinguished, why they could not interact with each other. Since the kinetic member of the (as we call them today, hypersymmetric (HySy)) field-charge pair depends not only on the local coordinates but also on velocity when it interacts with its potential field-charge sibling, it must do so in the presence of a velocity-dependent field (cf., Section 6). One could derive from these two assumptions the necessity of the existence of two conserved Noether currents. One of them coincides with what we knew "classically", the other includes a covariant derivative that hints at the gauge character of the velocity-dependent field (cf., Section 7). This latter current, due to its velocity dependence, appears only at higher velocities (higher energies), that means, it does not influence the predominance of the SM among "normal" circumstances (in other words, its effect is broken at everyday energies, cf., Sections 10.3.2–10.3.4).

The new conserved current hinted at a symmetry that conjectured a conserved property (this was the mentioned IFCS), and a mediating boson that transforms this property from one member of the HySy pair to the other (cf., Section 8 and Section 9.3).

One must premise another essential characteristic of the derivation: it did not presume any restriction for the Lagrangian in it (Section 7); consequently, its results could be applied to any kind of interaction. This was the basis of the conjecture that the derived conserved property, as well as the boson family mediating its transformation, may be interaction independent, universal, although the form of appearance of the given bosons should be individual in each kind of interaction.

In short, the breakthrough was brought about the conclusion that the originally *ad hoc* idea led to an invariance that involved the existence of a conserved property, and one could conclude a mediating particle (family) from this directly. At the turn of 2004–2005, it became clear that insofar nature verifies the mathematical conclusions, HySy may become an alternative of SUSY, as well as that the two together exclude each other.

Following the advice of the long earlier passed away Ne'eman, the mathematical derivation was published in the autumn of 2008 first online, and printed in a less known journal (Darvas, 2009) in January 2009. Then IJTP, which earlier rejected the publication, published the paper extended with the physical content. The basic physical article was published in 2011 (Darvas, 2011). In 2013–14, there appeared the application of the theory for the quantum electrodynamics (QED) in two parts in the same journal (Darvas, 2013b, 2014). Its application to the gravitational equations was written in a few published lectures (Darvas, 2012a, b, d, e, f; 2017c; see also in Sections 11.2 and 11.3).

In the opening plenary program of the Symmetry Festival 2013, following the lectures by G. 't Hooft and D. Horváth, who discussed the most actual theme that time, the finding of the Higgs boson, and spoke only tangentially on SUSY, the author read a lecture on the conservation of the IFCS in the electromagnetic interaction (Darvas, 2013a). (He spoke about it at smaller conferences even before.) Although it still has not been widely accepted as an alternative theory to the SUSY, it was already impossible to take no notice of its theoretically calculated results.

It happened in the summer of 2014 – when preparing a lecture (Darvas, 2015b) for a conference dealing with hypercomplex numbers and the physical applications of Finsler geometry – that the algebra and group with its matrix representation was found, which proved to be suitable to describe transformations of 3+1 type quantities. Since the properties of the Dirac ρ and Pauli σ matrices were used in the course of the presentation of the properties of that group, following the order of letters in the alphabet, it got the name τ (tau) algebra, and similarly, the relatives corresponding to Dirac's γ matrices were called δ matrices. That group proved to be isomorphic with the $SU(2)$ group, which shows the analogy of the IFCS to the *spin*, the *isotopic spin*, and the *weak isotopic spin* (Darvas, 2018b).

Many peer-reviewed papers, lectures read at international conferences have been published, and a few were in preparation until 2020, and many seminars and lectures were given for students at universities worldwide in this theme.

Its test in different interactions is still ahead, but there are various proposals on how to demonstrate the functioning of the model in the individual fundamental physical interactions (Darvas, 2016a, Section 3.1).

12.2 Summary of the findings in the HySy model

12.2.1 SUSY and HySy

The assumption of SUSY was based on string theory(ies). So did most alternative models too. The hypothesis of HySy embodies an essentially different paradigm. They all aim at serving as theoretical bases of a NP beyond the SM. Mathematics allows more correct conclusions – that all start from correct premises – than can be realised in nature. This does not mean that those theories that should be rejected after certain time were erroneous, only not all of them can be realised, only one of the alternatives proves itself to be the proper description of the experienced phenomena.

As it was cited in the introductory chapter, G. 't Hooft (2005, section 12) wrote the following on the expectations set up for a NP sought beyond the SM: "What is generally expected is either a new symmetry principle or possibly a new regime with an altogether different set of physical fields." The HySy model, whose silhouettes were drawn by that time, met both requirements. Even it did some more. SUSY (and most of its alternatives) became on and on more complicated during the years. Beyond that – while both traditional physics and its hoped new extension are arranged around *symmetry principles* – another principle, which can be called "*simplicity principle*", prevails in nature too. This means that when a structure can be realised in nature "simple" from among a few mathematically possible options, then "not the more complicated" one will be realised; provided that the simplest structure is not excluded by another condition, although that latter instance places the given structure *ab ovo* out of the reviewable options. At the moment, HySy offers a simpler model than any other model, mostly based on string theories, not excluding, of course, the possibility of the assumption mentioned at the end of the previous sentence. The HySy model involves the least number of free parameters among the alternatives.

The simplified SUSY model, which was reduced to five (according to other sources, four) free parameters, has aborted. Experiments have not confirmed it. The extended minimal supersymmetric standard model (MSSM) includes 124 (according to other sources, 104) free parameters. A part of physicists insisting still on SUSY is hunting for such MSSM SUSY particles. Probably, so many parametric theories can hardly be realised in nature. It would be inconsistent with the simplicity principle. See them in Section 12.2.15 to compare the complexity/simplicity of SUSY and HySy.

12.2.2 What are those isotopic field-charges?

As mentioned in Section 12.1, field-charges are the sources of the individual interaction fields. Several signs hinted that the field source appearing in the potential (scalar) component of the Hamiltonian of an object is not identical with the field source appearing in the kinetic (vector) component of that Hamiltonian. So, the gravitational and the inertial masses in mechanics and the Coulomb and Lorentz charges in electrodynamics differ to some extent. At the same time, considering the covariance principle, this distinction is not absolute. What field-charges look gravitational mass, Coulomb charge in one reference frame, present themselves as inertial mass, Lorentz charge seen from another. Therefore, the two different, but in their fundamental properties coinciding, field-charge types can be reckoned as isotopes of each other and shall continue to be measured in identical units.

No restriction was prescribed for the Lagrangian in the description of the interaction of IFC (Darvas, 2009; Darvas, 2011; Section 7). So, it can be substituted by any Lagrangian describing individually each physical interaction. Accordingly, the sources of any interaction field may occur in the concerned Lagrangian. On this basis, one can presume also the sources of the weak and the strong interactions to do so. If the derived two conserved Noether currents (Sections 7.3–7.5 and 8) apply also for them, they must have IFC too. The kinetic isotopic pairs of the colour charges of the strong interaction can be uneath imagined, thanks to their confinement, nevertheless, we have at least two reasons to not exclude the possibility of this. Once, they interact with each other, exchange gluons, so the second conserved Noether current should apply also for them; the other reason is that they can oscillate in the confined narrow space available for them and may configure a velocity-dependent field around themselves.

12.2.3 Why we have not featured them in our physical equations?

Until we had no experience with relativistic high speeds, it had been superfluous. Later, the interpretation of weak equivalence – that considered the equivalent field-charges identical – made apparently unnecessary to distinct them. Albeit, since the beginning of the 1910s (Mies, von Laue), then referring to them, de Broglie and others beginning in the early 1920s dealt with the distinction of the two kinds of masses (c.f.: literature to the *distinction between gravitational and inertial masses* in Darvas, 2016a). The distinction lies in their transformation rules. (Different views were composed on this issue. Moreover, we find in the literature various formulas for the quantity of moving mass till today.) We meet the same at Møller (1931) in respect of the electric charges in the early thirties. In the derivation of the Einstein equation, one can regularly find simplification that reduces the formulation of GTR to relativistically not too high velocities (e.g. 't Hooft, 2002), and the same appears also in the approximations in solutions of the equations, beginning with Schwarzschild. Later, Dirac also derived his – anyway Lorentz invariant – equation for QED emphasised as an approximation (Dirac, 1928). These approximations were later forgotten, only the final issues became presented as absolute in textbooks, then in citations and published applications. We are used to it in school that when the potential energy of a dropped body transforms into kinetic energy, then one can reduce the mass in the equation. It was justified at low velocities. Why not could one do so at those velocities observed in the classroom? The difference was negligible, but this is not the case always.

12.2.4 The question of localisation

The source of the scalar field – potential field-charge – is localised in space–time; the source of the vector field – kinetic field-charge – is velocity dependent, it is localised in the configuration space. Accordingly, the potential component of the Hamiltonian is space–time dependent, its kinetic component is velocity dependent. The individual velocities in the velocity field depend, of course, on space–time coordinates, which means the kinetic part of the Hamiltonian is an indirect function of the latter. Vectors in the velocity field can be projected into vectors in the configuration space (cf., Section 10.3.2) and back. In the course of the discussion of the IFC, there is expedient to handle the velocities separated from the momenta of the carriers of the velocities, since the momentum depends also on the mass of the carrier. Introducing the velocity dependence in the description brought along to conclude that the velocity space behaves as a gauge field. So, the description of the HySy differs from the way of discussion of the classical, Hamiltonian mechanics and the quantum theory that apply descriptions in the phase space. The reason of this difference is that the discussion of the IFC in velocity field demands to separate the velocity and mass that were coupled in the momentum (often only symbolically, abstracted from the velocity-mass product – e.g. the momentum of the light), in respect to the earlier observation that they transform in different fields (although the two fields appear connected in the theory of the HySy).

Nature makes a distinction between the potential (ψ_V) and kinetic (ψ_T) states of particles. Two particles which do not differ in any other property can be distinguished whether they are in potential or kinetic states. A single particle permanently changes its state from potential (ψ_V) to kinetic (ψ_T) and back (cf., Section 10.1).

12.2.5 Interaction between the different components of a Hamiltonian?

Interaction takes place between the potential state of a particle ($\daleth_V$) and the kinetic state of the other ($\daleth_T$), and *vice versa* (cf., Section 10.1).

In principle, one could imagine four mechanisms.

- *Probability model*, in which the wave function is with certain probability at the amplitude of an isotopic state, and, respectively, with another probability at the amplitude of the other isotopic state in a given moment, and these probabilities change in time.
- *Harmonic oscillator model*, in which the field-charge oscillates between two stable (potential and kinetic) states, and its energy oscillates also continually

between the potential and kinetic components of the Hamiltonian; observing a many-particle system, a part of the particles is in one of the states, the rest is in the other state at a given moment (entanglement), so in a measurement, one can observe both components of the Hamiltonian.

- *Flip-flop model*, in which the particle changes suddenly its state between the two extremes, and its state commutes continuously between the potential and the kinetic one.
- Finally the *mediating particle model*, in which the change between the two states, is ensured by the exchange of a mediating boson.

The HySy theory derived the forth, the *mediating particle model* (Darvas, 2011; Section 10.1.2) (on reasons explained later).

The presence of the conserved current allowed to derive the conjecture of a conserved quantity. This was the property called IFCS. IFCS is a property that can be interpreted in a gauge field – abstract compared to the known others. One derived and interpreted this gauge field with the introduction of the velocity field. IFCS was the conjectured conserved quantity, which transforms like a vector under rotations in the IFCS field. If one wants to stick to all that field theory states on conserved quantities, then one must assume the existence of a mediating boson in this gauge field. This mediating boson is called *dion* (after the field marked by **D**). These were demonstrated by the mathematical derivation, and this justified to accept the fourth model.

Interaction occurs by the exchange of two bosons: *one of them* was described in the SM and responds for the respective field-charge (ר), (i.e. the quantum of the SM field derived in Section 8.1, as a consequence of the respective conservation law and the related dynamical field equations); while *the other one*, the dion (δ) is responsible for the swap of the IFCS (Δ) (i.e. the quantum of the non-SM field derived in Section 8.2, as a consequence of the respective conservation law and the related field equations and equations of motion). Dion mediates between a scalar and a vector state, its only quantum number is Δ which can take two, $\pm\frac{1}{2}$ positions.

12.2.6 Why only opposite IFCS state particles can interact with each other?

The mathematical derivation had a before not mentioned premise, namely that always opposite IFCS state field-charges interact with each other. The HySy model assumed a gauge invariance, mathematically derived in Section 7, that makes a distinction between the two states. This gauge invariance is interpreted in an IFCS (Δ) gauge field (**D**), in the form of rotations of the $\boldsymbol{\Delta}$ vectors. It characterises a symmetry (named HySy) of nature not described earlier. Two separate SM fermions obey the Pauli principle even without this premise, they do not interact with each other provided that they are in the same quantum state. They should not be necessarily in different IFCS states. Yet, this premise was justified by the mechanism that even a single particle can (and it does) transform from one of its IFCS states to the other, and it can interact with "itself" (in fact, with its IFC sibling) only between "its" two quantum states.

12.2.7 3+1 type quantities in physics

There are numerous 3+1 type quantities in physics (space–time, four–momentum, vector+scalar potentials, four-currents, etc.). Except for the space–time, there appears in all of them the source of a field as a constituent, so that the kinetic isotopic member of the field-charge belongs to the vector-like component, and its potential isotopic member belongs to the scalar component. HySy describes that their IFCS state distinguishes them from each other, and they can be transformed into each other.

12.2.8 The transformation group of HySy

Since the subject is a conserved quantity, its transformation can be characterised by a group (Darvas, 2015b, c; Section 5). This group can be represented by the matrices of the τ (tau) algebra:

$$\tau_1 = \begin{bmatrix} 0 & 0 & 0 & 1 \\ 0 & 0 & 0 & 1 \\ 0 & 0 & 0 & 1 \\ 1 & 0 & 0 & 0 \end{bmatrix}, \quad \tau_2 = \begin{bmatrix} 0 & 0 & 0 & -i \\ 0 & 0 & 0 & -i \\ 0 & 0 & 0 & -i \\ i & 0 & 0 & 0 \end{bmatrix}, \quad \tau_3 = \begin{bmatrix} 1 & 0 & 0 & 0 \\ 1 & 0 & 0 & 0 \\ 1 & 0 & 0 & 0 \\ 0 & 0 & 0 & -1 \end{bmatrix}$$

$$\tau_1^2 = \tau_2^2 = \tau_3^2 = \begin{bmatrix} 1 & 0 & 0 & 0 \\ 1 & 0 & 0 & 0 \\ 1 & 0 & 0 & 0 \\ 0 & 0 & 0 & 1 \end{bmatrix} = \mathbf{E}, \quad \{\tau_i, \tau_j\} = 0, \quad [\tau_i, \tau_j] = 2i\tau_k$$

It has been demonstrated that the matrices of the τ algebra generate a group. This is the HySy group, which, despite the unusual structure of its representation, is isomorphic with the *SU*(2) group. Among others, this buttresses up the analogy of the IFCS with the spin, the isotopic spin and the weak isotopic spin, only it transforms in another abstract field.

12.2.9 Velocity-dependent field? Velocity-dependent quantities in physics

It is common to speak about velocity spaces, configuration space. Velocity dependent field is not a standard item of physics vocabularies. There are all the more velocity-dependent quantities. Such are the kinetic energy, vector currents, Lorentz force, Coriolis force, and the vector component of a Hamiltonian and so on. The Lorentz transformation describes emphasised the change of physical quantities depending on the velocity of their carriers, and we require invariance under this transformation for all physical theories. When one derived the conservation of the IFCS, started off from a velocity-dependent transformation that led to a second conserved Noether current, and that could be interpreted in a velocity-dependent gauge field:

$$J_\alpha^{(1)\nu}(\dot{x}) = \partial_\mu F_\alpha^{(1)\mu\nu}(\dot{x}) \qquad \partial_\nu J_\alpha^{(1)\nu} = 0$$

$$J_\alpha^{(2)\nu}(x) = \partial_\mu F_\alpha^{(2)\mu\nu}(x) \qquad \partial_\nu J_\alpha^{(2)\nu} = 0$$

written in detail:

$$J_\alpha^{(1)\nu}(\dot{x}) = ig\frac{\partial L}{\partial(\partial_\nu \varphi_k)}(T_\alpha)_{kl}\varphi_l(\dot{x})$$

$$\partial_\mu F_\alpha^{(1)\mu\nu}(\dot{x}) = ig\frac{\partial L}{\partial(\partial_\nu \varphi_k)}(T_\alpha)_{kl}\varphi_l(\dot{x})$$

$$J_\alpha^{(2)\nu}(x) = ig\left[\frac{\partial L}{\partial(\partial_\mu \varphi_k)}(T_\alpha)_{kl}\varphi_l(\dot{x})\Lambda_\mu^\nu - C^\gamma_{\alpha\beta} D_{\dot{\omega},\beta}(\dot{x})\Lambda_\mu^\omega \times F_\gamma^{(2)\mu\nu}(x)\right]$$

$$\hat{\partial}_\mu F_\alpha^{(2)\mu\nu}(x) = ig\frac{\partial L}{\partial(\partial_\mu \varphi_k)}(T_\alpha)_{kl}\varphi_l(\dot{x})\Lambda_\mu^\nu$$

where g is the coupling constant, T_α denotes the generator elements of the transformation group, φ the matter field, D the components of the velocity-dependent gauge field, C-s denote structure constants, $\hat{\partial}$ covariant derivative, and λ is an acceleration-like quantity that denotes the time derivatives of the Lorentz-boosted velocity components.

12.2.10 Conserved current – conserved quantity – mediating boson (dion)

The field-charge (ך) is a quantity conserved under the IFC transformation (cf., Section 8.1).

In a mathematical sense:

$J^{(1)}$ ensures the conservation of the field-charge (field source) known from the SM. The fact that the HySy theory complements the SM, but does not change it, is justified by the result that along with the introduction of new premises we have got $J^{(1)}$ in unchanged form.

$J^{(2)}$ ensures the conservation of another quantity, the IFCS (Δ). The full IFCS $\Delta = \frac{i}{g}\int J^{(2)4} d^3x$ is independent of time and is Lorentz invariant. $J^{(2)\mu}$ does not transform like a vector, but Δ does so under rotations in the IFCS field.

In a physical sense:

The two conserved currents (with restrictions detailed later) are present simultaneously. The derived conservation law demonstrates just the invariance of the two isotopic states of field-charges; a property that is independent of the nature of the concrete field whose sources are in subject. The source of IFCS (Δ) is the $\varphi(\dot{x})$ matter field, in interaction with the **D** kinetic gauge field.

There must exist quanta belonging to the **D** field. The quanta of the **D** field mediate the Δ IFCS. The Δ doublet as a conserved quantity is associated with the two isotopic states of the field-charges, and the related (Δ_i) operators induce the transfer from one member of the doublet to the other. These quanta are called δ bosons (dions).

The IFCS (Δ) invariance (cf., Section 8.2) concerns all types of particles that can occur in both potential (ψ_V) and kinetic (ψ_T) states. (All baryons, leptons, and unit mass are assumed to do so.) Δ appears as a property characterising the agents of various individual physical interactions. This statement needs additional remarks concerning the individual fundamental physical interactions. We presented examples picked up from SM interaction types in Sections 11.2–11.5.

12.2.11 Properties of the isotopic field-charges and the δ bosons

Both members of the IFC hyperpairs possess all properties that are known in the SM, and besides they possess the half-integer (±½) *IFCS* (Δ).

The properties of the *dions* (δ), as bosons: their *IFCS* (Δ) is an integer and they have *mass*. Their all other quantum numbers are 0 since their absorption and emission does not change the other properties of the particles. Their *IFCS* (Δ) is independent of the type of the interaction which they originate from, it is a property of the velocity-dependent gauge field.

12.2.12 How much is the mass of a dion?

Particles in the potential state are bare. Renormailised (dressed) mass and charge should be attributed to the kinetic states. Dions mediate between a particle in the kinetic IFC state with Lorentz-boosted (dressed) mass and a particle in the potential IFC state with rest mass. Consequently, a dion must transmit the *difference between the Lorentz transformed and the rest mass* from one to the other.

12.2.13 Massive bosons in the D field?

The presence of a massive mediating boson assumes a spontaneous symmetry breaking, which is compensated by a transformation in the respective gauge field. In the instance of the IFC field (**D**), the quanta of this field (δ) are associated with the conservation of the IFCS (Δ). The transformation that compensates the spontaneous symmetry breaking should depend on the velocity of the interacting IFC relative to each other. Consequently, in contrast to the SM bosons, the mass of the dions (δ bosons) has no fixed value (cf., Sections 10.3.2–10.3.4). The δ bosons never appear alone. They act simultaneously (parallel or antiparallel) with one of the SM bosons. Therefore, one requires the transformation of the **D** field together with one of the SM fields. The common rotation angle of the field **D** with a SM field is found $\varphi = \pi$. However, this angle $\varphi = \varphi(v, \theta_D)$ depends on the velocity of the field $\boldsymbol{v}$ and on a precession angle Θ_D of $\boldsymbol{v}$ around the effective velocity component $\boldsymbol{v}_3$ that latter arrows in the direction of Δ, which coincides with the direction of the velocity of the interacting IFC relative to each other (see Section 10.3.3, Figure 10.2). The Θ_D spontaneous precession angle characterises the transformation of the non-SM field **D** that gives mass to the boson δ and determines the transformation of the **D** field that compensates the spontaneous symmetry breaking, which is responsible for the mass of the field's δ boson (Darvas, 2020a-b). The relation $\varphi = \varphi(v, \theta_D)$ defines a speed (energy) limit for the observability of HySy: $\sqrt{8/9} \leq (v/c) < 1$ (cf., Section 10.3.3), while Θ_D may vary between $-(\pi/2) < \Theta_D < 0$.

12.2.14 How many kinds of dion are sought?

Mediating bosons belong to the sources of all interactions in the SM. The HySy model doubled these sources. New bosons, the dions, should mediate between these doubled source pairs. This means, we should conjecture as many dions as many bosons exist according to the SM. Besides, one dion hyperpair should be associated to the graviton in gravitation, like in the SM one to the photon in electromagnetics, three to the weak vector bosons, and eight to the gluons in the strong interaction. This makes altogether 13.

12.2.15 Why is HySy simpler than the SUSY model?

HySy is simpler than SUSY because we have to hunt for fewer particles. We do not need to find the split pairs of the field-charges, they have been here with us, only we have not attributed isolated attention to them. We should find (demonstrate) the 13 mediating bosons, whose mass can be calculated in each case separately.

12.2.16 Why is (if at all) HySy more complicated than the SUSY model?

All physical theories have been predicting fixed mass particles. We knew what mass to hunt for. The mass of the dions depends on the velocity of the interacting particles relative to each other. At high velocities, a small uncertainty in measuring the velocity may cause a large difference in the mass of the sought dion. We do not need enormous energies, but more precision in measuring.

12.2.17 Fermion–fermion and boson–boson pairs instead of fermion–boson pairs

Over the novelty that we are seeking for inconstant mass bosons, it is also new among the HySy paradigms of NP that we have fermion–fermion and boson–boson hyperpairs instead of fermion–boson and boson–fermion superpairs. The field-charges of the SM split into isotopic pairs in HySy. Between these fermion–fermion pairs new mediating bosons appear on the scene that act simultaneously and parallel (or antiparallel) with the respective SM mediating boson.

12.2.18 Do the two conserved Noether currents act together or separately?

The two conserved Noether currents are not completely independent of each other (cf., eqs. (7.4)–(7.7)). There is certain connectedness between them. At the same time, they also prevail independently. In principle, both are present always. However, at low energies – at low velocities – the effect of the second Noether current (7.24) is negligible, it does not prevail. The difference between the dressed and rest masses of a particle becomes significant at about $0.8c$. As we demonstrated in Section 10.3.3, HySy starts to affect at a critical real velocity $v = 0.943c$ (Figure 10.3) and, which is the same in another view, at an effective velocity $v_3 = 0.666c$ (Figure 10.4). Therefore, HySy is a "broken symmetry", whose effect does not prevail below the energy range defined by the critical value of v. This explains, why do most of our theories work without it, and this is the reason why we have not observed its effect in most experiences. On the other hand, at high kinetic energies, approximately above $0.998c$ speed, the rest mass becomes negligible compared to the dressed one. In experiments at those energies, the mass of the emitted dion practically coincides with the mass of the moving particle, in a reference frame fixed to the other particle or the lab. At those energies, we can disregard the effect of the first (that means, the classical, known in the SM) Noether current.

12.2.19 The SM and the HySy

HySy leaves the SM entirely alone. The effect of HySy remains latent up to not too much relativistic energies, similar to that of the electromagnetic field within the range of the strong interaction. HySy extends the SM, like relativistic and quantum physics extend Newtonian physics. Its effect strengthens with the increase of the kinetic energy. It does not influence the existing properties of the SM particles. At high energies, where HySy predominates, the group of the HySy extends the symmetry group of the individual interactions, and they prevail together. This concerns the Lorentz group as well [cf., e.g. eq. (10.10)]. The requirement that any physical theory should be Lorentz invariant is only a necessary condition, but it is not sufficient in all instances. In the presence of a velocity-dependent gauge field (that means, at high energies) one must extend the group of the Lorentz transformation with the group of the HySy, and investigate the invariance under the combined action of the two groups.

12.2.20 GUT and HySy

HySy may contribute to the GUT with a single, but powerful datum: it can be applied to any interaction, among others also to gravitational, thanks to the absence of restriction for the Lagrangian applied in its derivation (cf., Section 7.3).

12.2.21 Dark particles?

Probably, there exists not only one kind of dark matter in the universe. Probably, not a single theory gives an account on all of them. There happen, presumably, many high energy processes in the universe. There may occur in them – among others – HySy as well. Particles commute permanently their IFC states. At high energies, there is much mass just in the "bag of the postmen", that means, in its way between particles possessing kinetic and gravitational (potential) masses, and not only in course of the gravitational interaction. A part of the mass of these mediating dions may remain in dark for us.

12.2.22 Wave-corpuscle dualism

Particles being just in potential isotopic state behave like corpuscles. Those just in kinetic isotopic state behave like wavicles. Since they exchange dion between their states at high frequency, they appear sometimes as corpuscles, sometimes as waves for us, while they commute between their IFC states. The HySy theory may be suitable to explain the centuries-old dilemma of wave-corpuscle dualism.

Nevertheless, the properties that determine whether being a particle or wavicle are attributed not to physical objects but to their respective individual IFC. (That means, we have no reason to assume that an object's mass, electric charge, flavour, and colour are in the same isotopic Δ state simultaneously.)

12.2.23 Why does not the electron run away?

The HySy theory hides many further opportunities to explain phenomena that have been considered a paradox. For example, one of these is that the wave function of the electron – without interaction for long – should run away. It does not yet. Assuming that it permanently commutes between its two possible IFC states, one can imagine that it does not run away too wide, because it has not enough time to do so between two wave-corpuscle state changes. The frequent commutation between the two IFC states gives no opportunity for an interacting particle to spend a long time in one of the two Δ states. This circumstance does not allow the runaway growth, for example, for an electron cloud up to infinity, whose wave function is predicted to do so in a free kinetic state, because before its wavicle (and the surrounding thundercloud) would spread out too far, it will switch to its opposite (potential) Δ state. When it will switch back again to kinetic state, the runaway growth and building up an electric charge thundercloud starts again from its bare appearance.[12.1] An analogous process is supposed to take place with other $\daleth$–s, like mass and colour charge in a similar way as predicted by F. Wilczek (2003, p. 33). This oscillating mechanism may explain different behaviour of a fermion in certain aspects in bound and in free states.

12.2.24 Table of a few properties of isotopic field-charges

The two basic parts of a Hamiltonian can be characterised in the proposed model with the following dual properties:

Property	H_V	H_T
Energy	*Potential*	*Kinetic*
Field	*Matter field*	*Gauge field*
Localisation	*In space–time*	*In velocity field*
Sources	*Field-charges*	*Field-charge currents*
Transformation of the source	*Scalar*	*Vector*
Isotopic field-charge	$\daleth_V$	$\daleth_T$
State of the source	*Bound*	*Free*
Observable status	*Corpuscle*	*Wave*
Wave function component	*Longitudinal*	*Transversal*

12.1 Dirac (1962, p. 64) proposed also a possible model that eliminated the runaway motions for the electron.

12.3 Hypersymmetry and our picture of the physical world

HySy allowed us to look at the physical world through other eyes like we did it before.

This theory takes into consideration the transformation of the velocity and the respective field-charges separated. Therefore, in contrast to the phase space of classical quantum field theories, it applies description in the configuration space. Thus, a moment is separated into its velocity and field-charge components. These components transform in different fields. The latter has been introduced in the IFCS theory. This separation complies with a property of the Lorentz group ($SO^{+}(3,1)$) that splits into subgroups

The new invariance described in this book (and introduced in the cited previous papers in the theme) is interpreted in the presence of a kinetic gauge field. The Lorentz invariance is a combined symmetry in itself. The $SO^{+}(3,1)$ group of the proper Lorentz transformation can be characterised by two independent subgroups represented by six matrices. These two independent subgroups can be separated and characterised by three [4×4] rotation matrices [$R(\varphi)$] in the space–time, and three [4×4] velocity boosts [$\Lambda(\dot{x})$] into a given direction in the configuration space. They can be denoted together as $(= \Lambda \otimes R)$. Now, we should add to them the IFCS transformation: $\varDelta \otimes (\Lambda \otimes R)$. Since the added IFCS invariance [$\varDelta(\dot{x})$] is interpreted also in a kinetic field, it seems reasonable to substitute $(\varDelta \otimes \Lambda) \otimes R$ for the $\varDelta \otimes (\Lambda \otimes R)$ transformation. Although this clustering is only formal (due to the associativity of group operations), we must mention that the reason is to associate the velocity-dependent transformations (Λ and $\varDelta$) with each other, and formally separate them from the space–time rotation R. This re-clustering requires certain intendment change in the approach to the world picture.

$R(\varphi)$, $\Lambda(\dot{x}_v)$, and $\varDelta(\dot{x}_v)$ are universal invariance groups, that is, they concern all fundamental interactions. Nevertheless, the *IFCS*'s $SU(2)$ isomorphic symmetry [characterised by $\varDelta(\dot{x}_v)$] is broken at lower energies (cf., Section 10.3.3). This means, at not too high energies we get back to the $\Lambda \otimes R$ transformation used in the SM. At the same time, these invariances – both in SM and beyond the SM – are always extended by a specific invariance group, characteristic to the given interaction; for example, in case of electromagnetic interaction by $U(1)$, in case of electroweak interaction by $U(1) \otimes SU(2)$, and in case of strong interaction by $SU(3)$.

Introduction of the IFC distorted the covariance of our equations under Lorentz transformation, at least in the domain (energies) of HySy. We should insist on the covariance of our theories, but not just on their Lorentz covariance. The proven conservation of the IFCS demonstrated that the – apparently lost – covariance of our physical theories has been restored under the combined invariance under rotation of the IFCS in a kinetic gauge field plus Lorentz transformation together [$G \otimes SU(2)$, where G denotes the symmetry group of an arbitrary physical interaction in the SM].

In short, we showed that there exists an invariance group that can transform sources of 3+1 element physical quantities into each other. The group of the τ-matrices can make a unique correspondence between vector components and scalars. This group is isomorphic with the special unitary *SU*(2) group. This group can make a correspondence between such isotopic physical quantity siblings like the inertial and gravitational masses,[12.2] Lorentz type and Coulomb type electric charges, as predicted analytically (Darvas, 2011, 2013a-b, 2014)[12.3] and discussed here in Sections 7 and 10, earlier in Darvas (2017a), and can be applied also in the algebra of the genetic code (cf., Section 5.3.3). The group of the τ algebra defined an invariance between particles that composed HySy pairs. These particle twin siblings differ in their nature and physical properties from those predicted in the SUSY. For the sake of distinction, we called their invariance HySy.

12.4 Closing remarks

Steven Weinberg (1979) formulated in his Nobel lecture: "In the seventh book of the *Republic,* Plato describes prisoners who are chained in a cave and can see only shadows that things outside cast on the cave wall. [. . .] We are in such a cave, imprisoned by the limitations on the sorts of experiments we can do. In particular, we can study matter only at relatively low temperatures, where symmetries are likely to be spontaneously broken so that nature does not appear very simple or unified. We have not been able to get out of this cave, but by looking long and hard at the shadows on the cave wall, we can at least make out the shapes of symmetries, which though broken, are exact principles governing all phenomena, expressions of the beauty of the world outside."

Forty years have slid away, and "prisoners of the Standard Model" escaped from their cave. At the light of the real world, they became able to get convinced in experiences about the correctness of all their conclusions deduced from "shades on the cave's back wall". The SM has been completed. However, before that completion, we faced "facts" indicating limits of the SM. There appeared needs for new theories (called NP) to describe phenomena shown up beyond the SM – while not

---

12.2 Another approach to distinction between forms of masses is discussed (Calmet, Kuntz, 2017): According to them a few observed phenomena "suggest that there is a new form of matter that does not shine in the electromagnetic spectrum. Dark matter is not accounted for by either general relativity or the standard model of particle physics. While a large fraction of the high energy community is convinced that dark matter should be described by yet undiscovered new particles, it remains an open question whether this phenomenon requires a modification of the standard model or of general relativity. Here we want to raise a slightly different question namely whether the distinction between modified gravity or new particles is always clear". They showed that this is not always the case.

12.3 On the comparison of two descriptions of a quantum field theory, see Weiner (2011).

questioning the validity of the SM within its boundaries. "Prisoners of the NP" are still sitting in their cave back on the sally-gate attempting to reconstruct and make the real world clear based on limited "sorts of experiments" observed in shadow projections. Those shades suggested us shapes of new (super- or hyper-) symmetries (though broken within the usual limits) that govern phenomena beyond the SM. We hope, they indicate a no less beautiful world outside.

He closed his lecture (Weinberg, 1979) with the following words: "[Q]uantum field theory, which was born just fifty years ago from the marriage of quantum mechanics with relativity, is a beautiful but not very robust child. . . . at superhigh energies is susceptible to all sorts of diseases . . . and it needs special medicine to survive. One way that a quantum field theory can avoid these diseases is to be renormalisable and asymptotically free, but there are other possibilities. . . . Thus, one way or another, I think that quantum field theory is going to go on being very stubborn, refusing to allow us to describe all but a small number of possible worlds, among which, we hope, is ours." This book attempted "one way or another" to describe a predicted "possible" world.

REFERENCES

Abo-Zahhad, M., Ahmed, S. M., Abd-Elrahman, Sh. A. (2014) Integrated model of DNA sequence numerical representation and artificial neural network for human donor and acceptor sites prediction, Int. J. Inf. Technol. Comp. Sci. (IJITCS), 6(8), 51–57. https://doi.org/10.5815/ijitcs.2014.08.07.

Achiezer, A. I., Berestetskii, V. B. (1969) Kvantovaya elektrodynamika, 3rd modified edition, [in Russian] Moskva: Nauka, 623.

Ahmed, Z. (2003) Pseudo-reality and pseudo-adjointness of Hamiltonians, J. Phys. A: Math. Gen., 36(41), 10325. https://doi.org/10.1088/0305-4470/36/41/005.

Ahmed, Z., Jain, S. R. (2003a) Gaussian ensemble of 2 × 2 pseudo-Hermitian random matrices, J. Phys. A:Math. Theor., 36(12), 3349, https://doi.org/10.1088/0305-4470/36/12/327.

Ahmed, Z., Jain, S. R. (2003b) Pseudo-unitary symmetry and the Gaussian pseudo-unitary ensemble of random matrices, Phys. Rev. E, 67, 045106. http://arxiv.org/pdf/quant-ph/0209165.pdf.

Aichelburg, P. C. (2018) Symmetry principles in Einstein's theory of relativity, Symmetry: Cult. Sci., 2, 245–256. https://doi.org/10.26830/symmetry_2018_2_245.

Akhmedov, E. K., Smirnov, A. Y. (2011) Neutrino oscillations: Entanglement, energy-momentum conservation and QFT, Found. Phys., 41(8), 1279–1306. https://doi.org/10.1007/s10701-011-9545-4.

Aldea, N., Munteanu, G. (2015) A generalized Schrödinger equation via a complex Lagrangian of electrodynamics, J. Nonlinear. Math. Phy., 22(3), 361–373. https://doi.org/10.1080/14029251.2015.1056619.

Al-Kuwari, H. A., Taha, M. O. (1991) Noether's theorem and local gauge invariance, Am. J. Phys., 59 (4), 363–365. https://doi.org/10.1119/1.16551.

Araki, G., Huzinaga, S. (1951) Recoil effect on electron-proton forces and inapplicability of energy law, Prog. Theor. Phys., 6(5), 673–683. http://ptp.ipap.jp/link?PTP/6/673. https://doi.org/10.1143/PTP.6.673.

Barletta, E., Dragomir, S. (2012) Gravity as a Finslerian metric phenomenon, Found. Phys., 42(3), 436–453. https://doi.org/10.1007/s10701-011-9614-8.

Belavin, A., Polyakov, A., Schwartz, A., Tyupkin, Y. (1975), Phys. Lett., 59B, 85. https://doi.org/10.1016/0370-2693(75)90163-X.

Belot, G. (2003) Symmetry and Gauge Freedom, Studies in History and Philosophy of Science Part B: Studies in History and Philosophy of Modern Physics, 34(2), 189–225. https://doi.org/10.1016/S1355-2198(03)00004-2.

Bethe, H., Fermi, E. (1932) Über die Wechselwirkung von zwei Elektronen, Zeitschrift für Physik, 77 (5–6), 296–306. https://doi.org/10.1007/BF01348919.

Bethe, H. A., Bethe, H. (2002) Enrico Fermi in Rome, 1931–1932, Phys. Today, 6, 28. https://doi.org/10.1063/1.1496372.

Bohr, N. (1937), Philos. Sci., 4, 289. https://doi.org/10.1086/286465.

Bohr, N., Rosenfeld, L. (1950) Field and charge measurements in quantum electrodynamics, Phys. Rev., 78, 794. https://doi.org/10.1103/PhysRev.78.794.

Brading, K. A. (2002) Which symmetry? Noether, Weyl and conservation of electric charge, Stud. Hist. Philos. Mod. Phys., 33, 3–22. https://doi.org/10.1016/S1355-2198(01)00033-8.

Brading, K. A., Brown, H. R. (2000) Noether's theorems and gauge symmetries, http://arXiv:hep-th/0009058v1, 1–16.

Brading, K. A., Brown, H. R. (2003) Symmetries and Noether's theorems, In: Brading, K., Castellani, E. (eds.) Symmetries in Physics: Philosophical Reflections, 89–109, Cambridge: Cambridge University Press. 445. https://doi.org/10.1017/CBO9780511535369.006.

https://doi.org/10.1515/9783110713183-013

Brading, K. A., Brown, H. R. (2004) Are gauge symmetry transformations observable?, Br. J. Philos. Sci., 55(4), 645–665. https://doi.org/10.1093/bjps/55.4.645.

Breit, G. (1929), Phys. Rev., 34, 553. https://doi.org/10.1103/PhysRev.34.553.

Breit, G. (1932), Phys. Rev., 39, 616. https://doi.org/10.1103/PhysRev.39.616.

Brinzei, N., Siparov, S. (2008) Equations of electromagnetism in some special anisotropic spaces, arXiv:0812.1513v1 [gr-qc] 15 p.

Brody, D. C. (2014) Biorthogonal quantum mechanics, J. Phys. A: Math. Theor., 47(3), 035305. http://arxiv.org/pdf/1308.2609v2.pdf. https://doi.org/10.1088/1751-8113/47/3/035305.

Cabibbo, N. (1963) Unitary symmetry and leptonic decays, Phys. Rev. Lett., 10, 531. https://doi.org/10.1103/PhysRevLett.10.531.

Calmet, X., Kuntz, I. (2017) What is modified gravity and how to differentiate it from particle dark matter?, Eur. Phys. J. C, 77, 132. https://doi.org/10.1140/epjc/s10052-017-4695-y.

Castellani, E. (2001) Reductionism, emergence and effective field theories. arXiv:physics/0101039v1, 18 pp.

Castellani, E. (2003) Symmetry and equivalence, In: Brading, K., Castellani, E. (eds.) Symmetries in Physics: Philosophical Reflections, 425–436, Cambridge: Cambridge University Press. 445. https://doi.org/10.1017/CBO9780511535369.027.

Castellani, E. (2004) Dirac on gauges and constraints, Int. J. Theor. Phys., 43(6), 1503–1514. https://doi.org/10.1023/B:IJTP.0000048634.28339.24.

CERN workshop. (2005–2007) Report of Working Group 3 of the CERN Workshop "Flavour in the era of the LHC", Geneva, Switzerland, November 2005 – March 2007. http://arxiv.org/pdf/0801.1826v1.pdf.

CERN workshop (2008) Report, arXiv:0801.1800

CERN workshop (2008) Report, arXiv:0801.1833

CERN workshop (2008) Report, arXiv:0801.1826

Chanyal, B. C., Bisht, P. S., Negi, O. P. S. (2010) Generalized octonion electrodynamics, Int. J. Theor. Phys., 49(6), 1333–1343. https://doi.org/10.1007/s10773-010-0314-5.

Chartier, Émile-Auguste. (1938) Propos sur le Religion, 74.

Darvas, G. (2001) Symmetry and asymmetry in our surroundings; Aspects of symmetry in the phenomena of nature, physical laws, and human perception, In: Weibel, P. (ed.) Olafur Eliasson: Surroundings Surrounded, Essays on Space and Science, 136–149, Karlsruhe: Center for Arts and Media,. 703.

Darvas, G. (2002) Generalisation of the concept of symmetry and its classification in physics, The Official Electronic Proceedings Issue of the Wigner Centennial Conference, Pécs, 8–12 July, 2002, CD-ROM, item 48, 7 p.

Darvas, G. (2004) Generalisation of the concept of symmetry and its classification in physics, Acta Phys. Hungarica, 19(3), 373–379. https://doi.org/10.1556/APH.19.2004.3-4.39.

Darvas, G. (2007a) Symmetry, Basel: Birkhäuser. (2007), xi+508 p.

Darvas G. (2007b) A velocity dependent gauge invariance, In: Symmetry in Nonlinear Mathematical Physics, Kiev: Institute of Mathematics, Ukrainian Academy of Sciences. http://www.imath.kiev.ua/~appmath/Abstracts2007/Darvas.html.

Darvas, G. (2008) The Unreasonable Effectiveness of Symmetry in the Sciences, (Invited paper), Altenberg Workshop of Theoretical Biology September 11–14, In: Bookstein, F. L., Schaefer, K. (Eds) Measuring Biology – Quantitative Methods: Past and Future, Philosophical and Historical Foundations, Altenberg: Konrad Lorenz Institute for Evolution and Cognition Research, 14–15, https://www.kli.ac.at/webroot/files/file/Workshop%20booklets/19%20AWTB_Program%2BAbstracts.pdf.

Darvas, G. (2009) Conserved Noether currents, Utiyama's theory of invariant variation, and velocity dependence in local gauge invariance, Concepts Phys., VI(1), 3–16. https://doi.org/10.2478/

v10005-009-0001-6. http://arxiv.org/abs/0811.3189v1; http://www.hrpub.org/download/20040201/UJPA-18490279.pdf.
Darvas, G. (2011) The isotopic field charge spin assumption, Int. J. Theor. Phys., **50**(10), 2961–2991. https://doi.org/10.1007/s10773-011-0796-9 http://www.springerlink.com/content/g28q43v2112721r1/, http://www.springerlink.com/openurl.asp?genre=article&id=doi:10.1007/s10773-011-0796-9.
Darvas, G (2012a) Finsler geometry in GTR in the presence of a velocity dependent gauge field, Bull. Transilvania Univ. Brasov, Ser. III: Math. Info. Phys., 5(54), 2, 23–34.
Darvas, G (2012b) GTR and the isotopic field charge spin assumption, Hypercomplex Numbers Geom. Phys., 1(17), **9**, 50–59.
Darvas, G (2012c) Finslerian approach to the electromagnetic interaction in the presence of isotopic field-charges and a kinetic field, Hypercomplex Numbers Geom. Phys., 2(18), **9**, 1–19.
Darvas, G. (2012d) Isotopic Field Charge Spin Conservation in General Relativity Theory, pp. 53–65, In: M. C. Duffy, V. O. Gladyshev, A. N. Morozov, P. Rowlands (Ed.) Physical Interpretations of Relativity Theory, Moscow, Liverpool, Sunderland: Bauman Moscow State Technical University, 347.
Darvas, G. (2012e) Finsler geometry in the presence of isotopic field-charges applied for gravity, pp. 17–42 In: Proceedings of the Vth Petrov International Symposium, "High Energy Physics, cosmology and Gravity, Ed. S. Moskaliuk, Kiev: TIMPANI, 299 p.
Darvas, G. (2012f) Application of the isotopic field-charge assumption to the electromagnetic interaction, In: Tezisy dokladov konferentsii FERT – 2012 "Finslerovskie Obobshcheniya Teorii Otnositelnostyi", 25.06-1.07.2012.g., 48–49, Russia: Moskva-Fryazino-Lesnoe ozero. 101.
Darvas, G (2013a) A symmetric adventure beyond the Standard Model – Isotopic field-charge spin conservation in the electromagnetic interaction, Symmetry: Cult. Sci., 24(1–4), 17–40. https://doi.org/10.26830/symmetry_2013_1-4_017.
Darvas, G (2013b) The isotopic field-charge assumption applied to the electromagnetic interaction, Int. J. Theor. Phys. (2013) 52: 3853–3869https://doi.org/10.1007/s10773-013-1693-1http://www.springerlink.com/openurl.asp?genre=article&id=doi:10.1007/s10773-013-1693-1(online), 52(11), 2013, 3853–3869 (printed).
Darvas, G. (2014) Electromagnetic interaction in the presence of isotopic field-charges and a kinetic field, Int. J. Theor. Phys., 13 p. https://doi.org/10.1007/s10773-013-1781-2,. http://www.springerlink.com/openurl.asp?genre=article&id=doi:10.1007/s10773-013-1781-2 (online, October 2013), 53(1), 2014, 39–51 (printed).
Darvas, G. (2015a) The unreasonable effectiveness of symmetry in the sciences, Symmetry: Cult. Sci., 26(1), 039–082. https://doi.org/10.26830/symmetry_2015_1_039.
Darvas, G. (2015b) Quaternion-vector dual space algebras applied to the Dirac equation and its extensions, paper submitted to the X-th International Conference on Finsler extensions of relativity theory, August18–24, 2014 Braşov, Romania, Bull. Transilvania Univ. Brasov, Ser. III: Math. Info. Phys. 2015, 8(Issue 57–1), 27–42. https://www.researchgate.net/publication/283745150_Quaternionvector_dual_space_algebras_applied_to_the_dirac_equation_and_its_extensions.
Darvas, G. (2015c) Algebra of state transformations in strongly relativistic interactions, 13 p. https://www.researchgate.net/publication/295401749_Algebra_of_state_transformations_in_strongly_relativistic_interactions.
Darvas, G. (2015d) On Certain Questions Related to Information and Symmetries – In Physics From Certain View of Philosophy of Science, Paper submitted to the Conference: ISIS Summit Vienna 2015, Extended Abstract, 4 pages, Vienna-Basel: MDPI and ISIS, https://doi.org/10.3390/isis-summit-vienna-2015-T4002. https://www.researchgate.net/publication/300245902_On_

Certain_Questions_Related_to_Information_and_Symmetries_-_In_Physics_From_Certain_View_of_Philosophy_of_Science

Darvas, G. (2015e) How much relativistic was "classical" QED?, paper presented at the 2nd International Conference "Logic, Relativity and Beyond" August 9-13, 2015; in Book of Abstracts, pp. 6–8, Budapest, Alfréd Rényi Institute of Mathematics of the Hungarian Academy of Sciences, 20 p.

Darvas, G. (2016a) Isotopic field-charge (IFC) study project, https://www.researchgate.net/project/Isotopic-field-charge-IFC-study-project-G-Darvas.

Darvas, G. (2016b) The early years of QED (1928–1932) – in retrospect from the XXIst century, Paper submitted to the 2nd International Conference on the History of Physics, Europhysics, Pöllau, Austria, 5–7 September 2016, in: Europhysics Conference Abstract Volume 40C, p. 48, ISBN 979-10-96389-01-8, Mulhouse: European Physical Society, 60 p.

Darvas, G. (2017a) A few questions related to information and symmetries in physics, Eur. Phys. J. Special Topics, 226(2), 197–205. https://doi.org/10.1140/epjst/e2016-60356-1.

Darvas, G. (2017b) A few questions related to information and symmetries in physics: Unity through diversity, Information Studies and the Quest for Transdisciplinarity, https://doi.org/10.1142/9789813109001_0013.

Darvas, G. (2017c) Hypersymmetry of gravitational and inertial masses in relativistic field theories, 15 p; Paper presented at the XX International Conference „Physical Interpretations of Relativity Theory" – 2017, Bauman University, Moscow, 3–6 July 2017. Abstract: pp. 31–33, in: Physical Interpretations of Relativity Theory, Moscow, 3–6 July, 2017, Abstracts, Moscow: Bauman M. State Technical University, 164 p.

Darvas, G. (2017d) The nature of mass in logical perspective, paper presented at the 3rd International Conference "Logic, Relativity and Beyond" August 23–27, 2017; in Book of Abstracts, pp. 8–10, Budapest, Alfréd Rényi Institute of Mathematics of the Hungarian Academy of Sciences, 23 p.https://www.researchgate.net/publication/327232638_The_nature_of_mass_in_logical_perspective

Darvas, G. (2018a) Hypersymmetry as a New Paradigm in Contemporary Physical World-View, 11th International Symposium honouring noted mathematical physicist Jean-Pierre Vigier, In: „ADVANCES IN FUNDAMENTAL PHYSICS, Prelude to Paradigm Shift", 6–9 August 2018, Liege, Belgium, https://www.researchgate.net/publication/327232452_Hypersymmetry_as_a_New_Paradigm_in_Contemporary_Physical_World-View., http://www.noeticadvancedstudies.us/DarvasXI.pdf.

Darvas, G. (2018b, submitted) Algebra of hypersymmetry (extended version) applied to state transformations in strongly relativistic interactions illustrated on an extended form of the Dirac equation, preprint, 19 p.,https://arxiv.org/abs/1809.05396 https://www.researchgate.net/publication/327232210_Algebra_of_hypersymmetry_extended_version_applied_to_state_transformations_in_strongly_relativistic_interactions_illustrated_on_an_extended_form_of_the_Dirac_equation, https://doi.org/10.13140/RG.2.2.11959.57766

Darvas, G. (2018c) Variation principles, invariance principles, symmetry principles, laws of nature, Symmetry Cult. Sci., 29(4), 453–474. https://doi.org/10.26830/symmetry_2018_4_453.

Darvas, G. (2020a) Hypersymmetry field rotation angle, Chinese Journal of Physics 66, 776–786. https://doi.org/10.1016/j.cjph.2020.05.013

Darvas, G. (2020b) Giving mass to the mediating boson of HyperSymmetry by a field transformation applying Higgs mechanism beyond the Standard Model, *J. Phys.: Conf. Ser.* 1557 012002, 1–12. https://doi.org/10.1088/1742-6596/1557/1/012002; Paper presented at the XXI International Conference „Physical Interpretations of Relativity Theory" – 2019, Bauman State Technical University, Moscow, 1–5 July 2019, abstract: pp. 32–33, in: Physical

Interpretations of Relativity Theory, Moscow, 1–5 July, 2019, Abstracts, Moscow: Bauman M. State Technical University, 121 p.
Darvas, G., Farkas, F. T. (2006) An artist's works through the eyes of a physicist: Graphic illustration of particle symmetries, Leonardo, 39(1), 51–57. https://doi.org/10.1162/002409406775452195.
Darvas G., Farkas, F. T. (2008): Quantum Scent Dynamics (QSD): A new composite model of physical particles, http://arXiv:0803.2497v1 [hep-ph] 21 p.
Darvas, G., Petoukhov, S. V., (2017) Algebra that demonstrates similitude between transformation matrices of genetic codes and quantum electrodynamics, poster presentation to the workshop organised 27. 11.2017at the "Competence center for algorithmic and mathematical methods in biology, biotechnology and medicine" of the Hochschule Mannheim, 7p, (2017).
Darvas, G., Petoukhov, S. V. (2019) Algebra for transforming genetic codes based on matrices applied in quantum field theories; chapter 5 in: series Advances in Intelligent Systems and Computing, in: Hu, Zhengbing, Petoukhov, Sergey, He, Matthew (Eds) Advances in Artificial Systems for Medicine and Education II, Vol. 902, ISBN: 978-3-030-12081-8 Springer Nature.xv +793 p.
de Broglie, L. (1923) Note. Radiations – Ondes et quanta, Comptes Rendus, 177, 507–510.
de Broglie, L. (1925) Recherches sur la théorie des quanta, Ann. Phys., 10(III), 22–128. https://doi.org/10.1051/anphys/192510030022.
de Haas, E. P. J. (2004a) The combination of de Broglie's harmony of the phases and Mie's theory of gravity results in a principle of equivalence for quantum gravity, Ann. Fond. Louis Broglie, 29(4), 707–726.
de Haas, E. P. J. (2004b) A renewed theory of electrodynamics in the framework of a Dirac ether, In: Proc. P.I.R.T.-IX (London 2004), Liverpool: PD Publications, 95–123,.
de Haas, E. P. J. (2005) From Laue's stress-energy tensor to Maxwell's Equations and the implications for Einstein's GTR. In: Duffy, M. C., et al. (ed.) Proceedings of the Int. Conference "Physical Interpretation of Relativity Theory (PIRT-05)". Bauman Univ. Press, Moscow.
Deriglazov, A., Nersessian, A. (2014) Rigid particle revisited: Extrinsic curvature yields the Dirac equation, Phys.Lett., A378, 1224–1227. https://doi.org/10.1016/j.physleta.2014.02.034.
Dieks, D, Lubberdink, A. (2011) How classical particles emerge from the quantum world, Found. Phys., 41(6), 1051–1064. https://doi.org/10.1007/s10701-010-9515-2.
Dirac, P. A. M. (1928) The quantum theory of the electron, Proceedings of the Royal Society A: Mathematical, Physical and Engineering Sciences, 117, 610–624. https://doi.org/10.1098/rspa.1928.0023
Dirac, P. A. M. (1929) A theory of electrons and protons, Proceedings of the Royal Society of London, 126, 801, 360–365. https://doi.org/10.1098/rspa.1930.0013
Dirac, P. A. M. (1951a) A new classical theory of electrons, Proc. Roy. Soc. A, 209, 291–296. https://doi.org/10.1098/rspa.1951.0204.
Dirac, P. A. M. (1951b), Nature, 168, 906–907. https://doi.org/10.1038/168906a0.
Dirac, P. A. M. (1962) An extensible model of the electron, Proc. Roy. Soc. A, 268, 57–67. https://doi.org/10.1098/rspa.1962.0124.
Duffy, M. C., Gladyshev, V. O., Morozov, A-N., Rowlands, P. (eds.) (2012) Physical Interpretations of Relativity Theory, Proceedings of the International Scientific Meeting PIRT-2011, Moscow, Liverpool, Sunderland: Bauman Moscow State Technical University, 347.
Dyson, F. (2005) Hans Bethe and Quantum electrodynamics, Phys. Today, 58(10), 48–50. https://doi.org/10.1063/1.2138420.
Einstein, A. (1905a) Zur Elektrodynamik bewegter Körper, Ann. Phys., 17, 891. English translation: on the electrodynamics of moving bodies. In: The Principle of Relativity. Methuen and Co., London (1923).

Einstein, A. (1905b) Ist die Trägheit eines Körpers von seinem Energiegehalt abhängig?, Ann. Phys., 18, 639. English transl.: Does the inertia of a body depend upon its energy-content? In: The Principle of Relativity. Methuen and Co., London (1923).. https://doi.org/10.1002/andp.19053231314.

Einstein, A. (1915) Feldgleichungen der Gravitation, Preussische Akademie der Wissenschaften, Sitzungsberichte, 844–847.

Einstein, A. (1916) Grundlage der allgemeinen Relativitätstheorie, Annalen der Physik ser. 4, 49, 769–822. https://doi.org/10.1002/andp.19163540702.

Einstein, A. (1918) Phys. Z., 19, 165–166. https://doi.org/10.1175/1520-0493(1918)46<165c:N>2.0.CO;2.

Einstein, A. (1919) Spielen Gravitationsfelder im Aufbau der materiellen Elementarteilchen eine wesentliche Rolle?, In: Lorentz, H. A., Einstein, A., Minkowsky, H. et al. (eds.) Königlich Preussische Akademie der Wissenschaften, Physikalisch-matematische Klasse, Sitsungsberichte, 349–356, Translated as: Do gravitational fields play an essential part in the structure of elementary particles of matter? In, The Principle of Relativity, 189–198, Dover (1923).

Einstein, A. (1921) A brief outline of the development of the theory of relativity, Nature, 106(2677), 782–784. https://doi.org/10.1038/106782a0.

Einstein, A. (1935) Elementary derivation of the equivalence of mass and energy, Bull. Am. Math. Soc., 41, 223–230. https://doi.org/10.1090/S0002-9904-1935-06046-X.

Englert, F., Brout, R. (1964), Phys. Rev. Lett., 13, 321. https://doi.org/10.1103/PhysRevLett.13.321.

Eötvös, R. v. (1910) Verhandlungen der 16 Allgemeinen Konferenz der Internationalen Erdmessung, G. Reiner, Berlin, 319.

Eötvös, R. v., Pekár, D., Fekete, E. (1922) Annalen der Physik 68, 11.

Fabbri, L. (2013) A discussion on Dirac field theory, no-go theorems and renormalizability, Int. J. Theor. Phys., 52(2), 634–643. https://doi.org/10.1007/s10773-012-1370-9.

Fermi, E. (1931) Le masse elettromagnetiche nella elettrodinamica. (Italian) [The electromagnetic mass in quantum electrodynamics], Nuovo Cimento 8, 1, 121–132.

Fermi, E. (1932) Quantum Theory of Radiation, Rev. Mod. Phys., 4, 87–132. https://doi.org/10.1103/RevModPhys.4.87.

Feynman, R. P. (1949) Space-time approach to quantum electrodynamics, Phys. Rev., 76(6), 769–789. https://doi.org/10.1103/PhysRev.76.769.

Fillion-Gourdeau, F., Herrmann, H. J., Mendoza, M., Palpacelli, S, Succi, S. (2014) Formal analogy between the Dirac equation in its majorana form and the discrete-velocity version of the Boltzmann kinetic equation, Phys. Rev. Lett., 111, 160602. http://dx.doi.org/10.1103/PhysRevLett.111.160602.

Gell-Mann, M., Ne'eman, Y. (2000) The Eightfold Way original edition, Benjamin, Elmsford: Re-published by Westview Press (1964), 336.

Gershon, T. (2008) Quantum loop effects, Phys. World, June 22–25. http://lhcb-public.web.cern.ch/lhcb/Hot%20News/Articles/PWJun08gershon.pdf.

Glashow, Sh. L. (1961), Nucl. Phys., 22, 579. https://doi.org/10.1016/0029-5582(61)90469-2.

Glashow, Sh. L.and Weinberg, S. (1977) Natural conservation laws for neutral currents, Phys. Rev. D, 15, 1958–1965. https://doi.org/10.1103/PhysRevD.15.1958.

Goenner, H. (2013) Weak Lie symmetry and extended Lie algebra, J. Math. Phys., 54, 041701. http://doi.org/10.1063/1.4795839.

Goldstone, J. (1961), Nuovo Cimento, 19,, 154. https://doi.org/10.1007/BF02812722.

Goldstone, J., Salam, A., Weinberg, S. (1962) Broken symmetries, Phys. Rev., 127(3), 965–970. https://doi.org/10.1103/PhysRev.127.965.

Gould, L. I. (1989) Nonlocal conserved quantities, balance laws, and equations of motion, Int. J. Theor. Phys., 28, 335–363. https://doi.org/10.1007/BF00670207.
Heisenberg, W. (1931), Ann. d. Phys., 9, 338. https://doi.org/10.1002/andp.19314010305.
Higgs, P. W. (1964) Broken symmetries and the masses of gauge bosons, Phys. Rev. Lett., 13(16), 508–509. https://doi.org/10.1103/PhysRevLett.13.508.
Higgs, P. W. (1966) Spontaneous symmetry breakdown without massless bosons, Phys. Rev., 145, 1156–1162. https://doi.org/10.1103/PhysRev.145.1156.
Hiley, B. J., Callaghan, R. E. (2012) Clifford Algebras and the Dirac-Bohm Quantum Hamilton-Jacobi Equation, Found. Phys., 42(1), 192–208. http://dx.doi.org/10.1007/s10701-011-9558-z.
Horváth, D. (2005) The deepest symmetries of nature: CPT and SUSY. Invited paper presented at Workshop on Physics with Ultra Slow Antiproton Beams, RIKEN, Wako, Japan, 14–16 March. http://www. kfki.hu/~horvath/RIPNP-GRID/Publications/2005/HD_RikenCPT05paper.pdf https://doi.org/10.1063/1.2121970
Horváth, D. (2006) Symmetries and their violation in particle physics, Symmetry: Cult. Sci., 17(1–2), 159–174. https://doi.org/10.26830/symmetry_2016_1-2_159.
Horváth, D. (2013) Hunting the Higgs-boson – Is it found at LHC?, Symmetry Cult. Sci., 24(1–4), 9–16. https://doi.org/10.26830/symmetry_2013_1-4_009.
Hraskó, P. (2001) Realativity Theory, [in Hungarian and English], Budapest: Typotex, 436.
Hraskó, P. (2003) Ekvivalens-e egymással a tömeg és az energia? [Are mass and energy equivalent? in Hungarian], Fiz. Szle., 53(9), 330.
Itzykson, C., Zuber, J. B. (1980) Quantum Field Theory, New York: McGraw-Hill.
Jackiw, R., Rebbi, C. (1976) Conformal properties of Yang-Mills pseudoparticle, Phys. Rev. D, 14(2), 517–523. https://doi.org/10.1103/PhysRevD.14.517.
Jentschura, U. D. (2013) Gravitationally coupled Dirac equation for antimatter, Phys. Rev. A, 87, 032101 and Phys. Rev. A 87, 069903(E. http://doi.org/10.1103/PhysRevA.87.032101.
Jentschura U. D., C. M. Adhikari (2018) Relativistic and radiative corrections to the dynamic stark shift: Gauge invariance and transition currents in the velocity gauge, June 2018, Phys. Rev. A, 97(6), 062120. https://doi.org/10.1103/PhysRevA.97.062120.
Julia, B., Zee, A. (1975) Poles with both magnetic and electric charges in non-Abelian gauge theory, Phys. Rev. D, 11, 2227–2232. https://doi.org/10.1103/PhysRevD.11.2227.
Kerner, R., Abramov, V., Le Roy, B. (1997) Hypersymmetry: A Z_3-graded generalization of supersymmetry, J. Math. Phys., 38, 1650. https://doi.org/10.1063/1.531821.
Kerner, R. (2018) Quantum Physical Origin of Lorentz Transformations, J. Phys. Conf. Series, 1051, (2018) 012018, https://doi.org/10.1088/1742-6596/1051/1/012018.
Landau, L. D., Lifshitz, E. M. (1967) Teoriya Polya, [Field Theory, In Russian], Nauka: Moskva, 458.
Martin, C. A. (2003) On continuous symmetries and the foundations of modern physics, In: Brading, K., Castellani, E. (eds.) Symmetries in Physics: Philosophical Reflections, 29–60, Cambridge: Cambridge University Press. 445. https://doi.org/10.1017/CBO9780511535369.004.
Meister, S., Stockburger, J. T, Schmidt, R., Ankerhold, J. (2014) Optimal control theory with arbitrary superpositions of waveforms, J. Phys. A: Math. Theor., 47, 49. https://doi.org/10.1088/1751-8113/47/49/495002.
Mie, G. (1912a), Ann. Phys., 37, 511–534. https://doi.org/10.1002/andp.19123420306.
Mie, G. (1912b) Grundlagen einer Theorie der Materie, Ann. Phys., 39, 1–40. https://doi.org/10.1002/andp.19123441102.
Mie, G. (1913), Ann. Phys., 40, 1–66. https://doi.org/10.1002/andp.19123441102.
Mills, R. (1989) Gauge fields, Am. J. Phys., 57(6), 493–507. https://doi.org/10.1119/1.15984.
Modified Gravity Theories. (2012–2014) Int. J. Mod. Phys. D. Selected papers published by the World Scientific Publishing Company.

Mostafazadeh, A. (2002) Pseudo-Hermiticity versus PT-Symmetry: The necessary condition for thereality of the spectrum of a non-Hermitian Hamiltonian, J. Math. Phys., 43, 205–214. math-ph/0107001.. https://doi.org/10.1063/1.1418246.

Mostafazadeh, A. (2004) Pseudo-unitary operators and pseudo-unitary quantum dynamics, J. Math. Phys., 45, 932–946. math-ph/0302050.. https://doi.org/10.1063/1.1646448.

Mostafazadeh, A. (2006) A physical realization of the generalized PT-, C-, and CPT-Symmetries and the position operator for Klein-Gordon fields, Int. J. Mod. Phys. A, 21, 2553–2572. quant-ph/0307059. https://doi.org/10.1142/S0217751X06028813.

Mostafazadeh, A. (2013) Pseudo-Hermitian Quantum Mechanics with Unbounded Metric Operators, Phil.Trans. R. Soc. A, 371, 20120050. 7 pages; arXiv: 1203.6241. https://doi.org/10.1098/rsta.2012.0050.

Mostafazadeh, A. (2014) Generalized unitarity and reciprocity relations for PT-symmetric scatteringpotentials, J. Phys. A: Math. Theor., 47, 505303. 6 pages; arXiv: 1405.4212. https://doi.org/10.1088/1751-8113/47/50/505303.

Munteanu, G. (2010) A Yang-Mills electrodynamics theory on the holomorphic tangent bundle, J. Nonlinear Math. Phys., 17(2), 227–242. https://doi.org/10.1142/S1402925110000738.

Myrzakulov, R., Sebastiani, L., Zerbini, S. (2013) Some aspects of generalized modified gravity models, Int. J. Mod. Phys. D, 22(8), 1330017, 172, https://doi.org/10.1142/S0218271813300176.

Møller, C. (1931) Über den Stoß zweier Teilchen unter Berücksichtigung der Retardation der Kräfte, Zeitschrift für Physik, 70(11–12), 786–795. https://doi.org/10.1007/BF01340621.

Neretin, Yu. A. (2011) Lectures on Gaussian integral operators and classical groups, Zürich: European Mathematical Society, 559 p. https://doi.org/10.4171/080

Ne'eman, Y. (1990) The interplay of symmetry, order and information in physics and the impact of gauge symmetry on algebraic topology, Symmetry: Cult. Sci., 1(3), 229–255. https://doi.org/10.26830/symmetry_1990_3_229.

Ne'eman, Y., Kirsh, Y. (1986) The Particle Hunters, Cambridge: Cambridge University Press, 272.

Noble, J. H., Jentschura, U. D. (2015a) Dirac equations with confining potentials, Int. J. Mod. Phys. A, 30, 1550002. e-print arXiv:1410.1516 [quant-ph]. https://doi.org/10.1142/S0217751X15500025.

Noble, J. H., Jentschura, U. D. (2015b) Ultrarelativistic decoupling transformation for generalized Dirac equations, Phys. Rev. A, 92, 012101. e-print arXiv:1506.02504 [gr-qc]. https://doi.org/10.1103/PhysRevA.92.012101.

Noether, E. A. (1918) Invariante Variationsprobleme, Nachrichten von der Königlichen Gesellschaft der Wissenschaften zu Göttingen: Mathematisch-physikalische Klasse, 235–257.

Norton, J. D. (2003) General covariance, gauge theories and the Kretschmann Objection, In: Brading, K., Castellani, E. (eds.) Symmetries in Physics: Philosophical Reflections, 110–123, Cambridge: Cambridge University Press. 445 p. https://doi.org/10.1017/CBO9780511535369.007.

Onishchik, A. L. Sulanke, R. (2006) Projective and Cayley-Klein Geometries, Springer Science & Business Media, 450.

Pauli, W. (1941) Relativistic field theories, Rev. Mod. Phys., 13, 203–232. https://doi.org/10.1103/RevModPhys.13.203.

Petoukhov, S. (2006) Bioinformatics: Matrix genetics, algebras of the genetic code and biological harmony, Symmetry: Cult. Sci., 17(1–4), 253–291. https://doi.org/10.26830/symmetry_2006_1-4_253.

Petoukhov, S. V. (2008) Matrix genetics, algebras of the genetic code, noise immunity, Moscow: RCD, [in Russian], 316.

Petoukhov S. V. (2011) Matrix genetics and algebraic properties of the multi-level system of genetic alphabets, Neuroquantology, 9(4), 60–81. http://www.neuroquantology.com/index.php/journal/article/view/501. https://doi.org/10.14704/nq.2011.9.4.501.
Petoukhov S. V. (2012) Symmetries of the genetic code, hypercomplex numbers and genetic matrices with internal complementarities, Symmetries in genetic information and algebraic biology, special issue Symmetry: Culture and Science, Guest editor: S. V. Petoukhov, 23(3–4), 275–301. https://doi.org/10.26830/symmetry_2012_3-4_275.
Petoukhov S. V. (2015a) The genetic code, 8-dimensional hypercomplex numbers and dyadic shifts, http://arxiv.org/abs/1102.3596, version 8, (2015), 83 pages, received on 18. 04.2015.
Petoukhov S. V. (2015b) Resonances and genetic biomechanics, Symmetry: Cult. Sci., 26(3), 379–397. https://doi.org/10.26830/symmetry_2015_3_379.
Petoukhov, S. (2017) Genetic coding and united-hypercomplex systems in the models of algebraic biology, Biosystems, 158, 31–46. August. https://doi.org/10.1016/j.biosystems.2017.05.002.
Petoukhov, S. (2018) The genetic coding system and unitary matrices, Preprints, 2018040131. https://doi.org/10.20944/preprints201804.0131.v1.
Podolsky, B. (1931) A tensor form of Dirac's Equation, Phys. Rev., 37, 1398. http://dx.doi.org/10.1103/PhysRev.37.1398.
Pons, J. M., Salisbury, D. C., Shepley, L. C. (2000) Gauge transformations in Einstein-Yang-Mills theories, J. Math. Phys, 41, 5557–5571. https://arXiv:gr-qc/9912086v1. https://doi.org/10.1063/1.533425.
Rabinowitz, M. (2013) Challenges to Bohr's wave-particle complementarity principle, Int. J. Theor. Phys., 52(2), 668–678. https://doi.org/10.1007/s10773-012-1374-5.
Rosen, J. (1995) Symmetry in Science, An Introduction to the General Theory, Berlin: Springer, 213 p.
Rosen, J. (2008) Symmetry Rules, How Science and Nature Are Founded on Symmetry, The Frontiers Collection, Berlin: Springer, 304, https://doi.org/10.1007/978-3-540-75973-7.
Rowlands, P. (2012a) A null Berwald-Moor metric in nilpotent spinor space, Symmetry: Cult. Sci., 23(2), 179–188. https://doi.org/10.26830/symmetry_2012_2_179.
Rowlands, P. (2012b) The null Berwald-Moor metric and the nilpotent wavefunction, 8th FERT.
Rowlands, P. (2013a) Symmetry in physics from the foundations, Symmetry: Cult. Sci., 24(1–4), 41–56. https://doi.org/10.26830/symmetry_2013_1-4_041.
Rowlands, P. (2013b) Dual spaces, particle singularities and quartic geometry, 9th FERT, 10p.
Rowlands, P. (2015a) How Schrödinger's Cat Escaped the Box, New Jersey, etc.: World Scientific, x + 187, ISBN 978-981-4635-19-6.
Rowlands, P. (2015b) The Foundations of Physical Law, New Jersey, etc.: World Scientific. 2015, xiv + 247, ISBN 978-981-4618-37-3.
Ruffini, R., Vereshchagin, G., Xue, S.-S. (2010) Electron–positron pairs in physics and astrophysics: From heavy nuclei to black holes, Phys. Rep., 487(1–4), 1–140. https://doi.org/10.1016/j.physrep.2009.10.004.
Salpeter, E. E., Bethe, H. A. (1951) A relativistic equation for bound-state problems, Phys. Rev., 84 (6), 1232–1242. https://doi.org/10.1103/PhysRev.84.1232.
Sarkar, K., Sk, N., Debnath, S., Sanyal, A. K. (2013) Viability of Noether symmetry of F(R) theory of gravity, Int. J. Theor. Phys., 52(4), 1194–1213. https://doi.org/10.1007/s10773-012-1436-8.
Schroer, B. (2011) An alternative to the Gauge theoretic setting, Found. Phys., 41(10), 1543–1568. https://doi.org/10.1007/s10701-011-9567-y.
Schweber, S. S. (2002) Enrico Fermi and quantum electrodynamics, 1929-1932, Phys. Today, 6, 31. https://doi.org/10.1063/1.1496373.
Schwinger, J. (1969) A magnetic model of matter, Science, 165(3895), 757–761. https://doi.org/10.1126/science.165.3895.757.

Singh, V., Tripathi, B. V. (2013) Topological Dyons, Int. J. Theor. Phys., 52(2), 604–611. https://doi.org/10.1007/s10773-012-1366-5.
Siparov, S. (2012) Introduction to the Anisotropic Geometrodynamics, Singapore: World Scientific, 303 p. https://doi.org/10.1142/9789814340847.
Slepyan, L. I. (2015) On the energy partition in oscillations and waves, Proc. R. Soc. A, 471, 2175. https://doi.org/10.1098/rspa.
Szent-Györgyi, A. (1985) in: Bridging the present and the future: IEEE Professional Communication Society conference record, Williamsburg, Virginia, October 16–18, p. 14.
't Hooft, G. (1974) Magnetic monopoles in unified gauge theories, Nucl. Phys. B, 79, 276–284. https://doi.org/10.1016/0550-3213(74)90486-6.
't Hooft, G. (2002) Introduction to General Relativity, Princeton: Rinton Press Inc.
't Hooft, G. (2005) The conceptual basis of quantum field theory, In: Handbook of the Philosophy of Science, Amsterdam: Elsevier.
't Hooft, G. (2011) A class of elementary particle models without any adjustable real parameters, Found. Phys., 41(12), 1829–1856. https://doi.org/10.1007/s10701-011-9586-8.
The Belle Collaboration. (2008) Difference in direct charge-parity violation between charged and neutral B meson decays, Nature, 452, 332–335, https://doi.org/10.1038/nature06827.
Utiyama, R. (1956) Invariant theoretical interpretation of interaction, Phys. Rev., 101(5), 1597–1607. https://doi.org/10.1103/PhysRev.101.1597.
Utiyama, R. (1959) Theory of invariant variation and the generalised canonical dynamics, Progr. of Theor. Phys. Suppl., 9, 19–44. https://doi.org/10.1143/PTPS.9.19.
Vaccaro, J. A. (2016) Quantum asymmetry between time and space, Proc. R. Soc. A, 472, 2185. https://doi.org/10.1098/rspa.2015.0670.
Vitiello, G., Blasone, M. (2004) Quantum Field Theory of Particle Mixing and Oscillations, In: Gruber B. J., Marmo G., Yoshinaga N. (eds) Symmetries in Science XI, xxiv + 612, Dordrecht: Kluwer Academic Publishers. 105–128. https://doi.org/10.1007/1-4020-2634-X_7.
Voicu, N. (2010) Equations of electromagnetism in some special anisotropic spaces, Part 2, Hypercomplex Numbers Geom. Phys., 7, 2(14), 61–72.
von Laue, M. (1911), Ann. Phys, 35, 524–542. https://doi.org/10.1002/andp.19113400808.
von Laue, M. (1955) Relativitätstheorie, 6th, Braunschweig.
Wani, A. A., Badshah, V. H. (2017) On the relations between Lucas Sequence and Fibonacci-like sequence by matrix methods, Int. J. Math. Sci. Comp. (IJMSC), 3(4), 20–36. https://doi.org/10.5815/ijmsc.2017.04.03.
Weinberg, S. (1967) A model of leptons, Phys. Rev. Lett., 19(21), 1264–1266. https://doi.org/10.1103/PhysRevLett.19.1264.
Weinberg, S. (1972) Effects of a neutral intermediate boson in semileptonic processes, Phys. Rev. D., 5(6), 1412–1417. https://doi.org/10.1103/PhysRevD.5.1412.
Weinberg, S. (1973) General theory of broken symmetries, Phys. Rev. D, 7, 1068–1082. https://doi.org/10.1103/PhysRevD.7.1068.
Weinberg, S. (1976) Implications of dynamical symmetry breaking, Phys. Rev. D., 13(4), 974. https://doi.org/10.1103/PhysRevD.13.974.
Weinberg, S. (1979) Conceptual foundations of the unified theory of weak and electromagnetic interactions, in: Lundqvist, S. (ed.) Nobel lectures, physics, 1971–1980, pp. 543–559, Singapore: World Scientific (1992).
Weinberg, S. (1980) Unified theory of weak and electromagnetic interactions (Nobel lecture), Rev. Mod. Phys., 52(3), 315. https://doi.org/10.1103/RevModPhys.52.515.
Weinberg, S. (1995) Foundations. The Quantum Theory of Fields, Vol. I., Cambridge: Cambridge University Press. https://doi.org/10.1017/CBO9781139644167.

Weinberg, S. (1996) Modern Applications. The Quantum Theory of Fields, Vol. II., Cambridge: Cambridge University Press. Chap. 15. Non-Abelian gauge theories.

Weinberg, S. (1997) Changing attitudes and the standard model, In: Hoddeson, L., Brown, L., Riordan, M., Dresden, M. (eds.) The Rise of the Standard Model, 36–44, Cambridge: Cambridge University Press. 38. https://doi.org/10.1017/CBO9780511471094.004.

Weinberg, S. (2011) Symmetry: A 'Key to Nature's Secrets', The New York Review of Books, October 27, http://www.nybooks.com/articles/archives/2011/oct/27/symmetry-key-natures-secrets/.

Weinberg, S. (2012) Varieties of symmetry, Symmetry: Cult. Sci., 23(1), 5–16. 41. https://doi.org/10.26830/symmetry_2012_1_005.

Weyl, H. (1918) Gravitation und Elektrizität, Sitzungsberichte Preussische Akademie der Wissenschaften, pp. 465–480.

Weyl, H. (1928) The Theory of Groups and Quantum Mechanics, New York: Dover.

Weyl, H. (1929) Electron and gravitation, In: O'Raifeartaigh (ed.) The Dawning of Gauge Theory, Princeton: Princeton Univ. Press.

Weyl, H. (1949) Philosophy of Mathematics and Natural Science, Princeton: Princeton University Press, 170–180. with a new introduction by F. Wilczek (2009) 315.

Weiner, M. (2011) An algebraic version of Haag's theorem, Commun. Math. Phys., 305(2), 469–485. https://doi.org/10.1007/s00220-011-1236-7.

Wigner, E. P. (1960) The unreasonable effectiveness of mathematics in the natural sciences, Commun. Pure Appl. Math., 13(1), February, 1–14, https://doi.org/10.1002/cpa.3160130102.

Wigner, E. P. (1967) Events, laws of nature and invariance principles, In: Wigner (ed.) Symmetries and Reflections, pp. 38–50. Bloomington: Indiana University Press.

Wigner, E. P. (1984) The meaning of symmetry, in: Zichichi, A. (ed.) Gauge Interactions Theory and Experiment, Proceedings of the 20th International School of Subnuclear Physics, Erice, Sicily, pp. 729–733, New York: Plenum.

Wilczek, F. (1998) Riemann-Einstein Structure from volume and Gauge symmetry, IASSNS-HEP-97/142, http://arxiv.org/pdf/hepth/9801184v4, http://arXiv:hep-th/9801184v4

Wilczek, F. (2003) The origin of mass, MIT Phys. Annu., 2003, 24–35. https://doi.org/10.1016/S0041-1345(03)00239-2.

Wilczek, F. (2008) Particle physics: Mass by numbers, Nature, 456, 449–450. https://doi.org/10.1038/456449a.

Wilczek, F. (2018) Has elegance betrayed physics?, Phys. Today, 71(9), 57–58. https://doi.org/10.1063/PT.3.4022.

Wüthrich, Ch., Huggett, N., Vistarini, T. (2012), Time in quantum gravity, http://arXiv:1207.1635v1, 18 p.

Yang, C. N., Mills, R. L. (1954) Conservation of isotopic spin and isotopic gauge invariance, Phys. Rev., **96**(1), 191–195. https://doi.org/10.1103/PhysRev.96.191.

Zee, A. (1999) Fearful Symmetry, extended edn., Princeton: Princeton University Press, 336 p.

INDEX

3+1 type quantities 49, 51, 55, 187, 192, 200

affine connection field 145
algebra of HySy / HySy algebra 54, 69, 112, 147, 157, 159
asymmetry 3, 31–33, 45, 112, 114–115, 128, 141, 152, 154–155, 163, 174, 180, 185

baryon 17–18, 99–100, 194
BEH mechanism 119, 128, 134
Bethe Ansatz 114–115
beyond the SM 3, 5–6, 7, 8, 81, 91, 101–102, 105, 119, 130, 134–135, 171, 175, 177–178, 181, 183, 186, 188, 191, 193–197, 199–201
boson 5, 8, 17, 24, 28, 81, 104–105, 111–112, 119–121, 125–126, 128, 130, 132–135, 137, 145–146, 163, 177–179, 181, 183, 186–187, 191, 193–196
bound state 31, 33, 44, 99–100, 109, 112–113, 155, 170
broken symmetry / symmetry breaking 6, 81, 105, 135, 179, 196

Cabibbo angle, CKM angle 120, 126
Clifford algebra 54, 147
colour 18–19, 20, 21, 43, 74, 99–100, 103, 110, 115, 182–183, 189, 197–198
commutation 52, 57, 85, 93, 96, 113, 150, 156, 164, 170, 172, 176, 180, 198
commutator, commute 53–54, 60, 96, 111–113, 149–150, 156, 164–165, 169–170, 172, 174, 176, 183, 191, 197–198
configuration space 75, 82–83, 94, 121–124, 127, 131–132, 134, 175–176, 190, 192, 199
connection field 172
conservation 6, 13–14, 24, 26–29, 48, 81, 93, 95, 97, 99, 101, 115–116, 137–138, 175, 177, 193
conservation law 28–29, 48, 75, 77, 81, 94, 100–101, 103, 154, 177, 191, 194
conservation of the isotopic field-charge spin 93, 97, 99, 116, 121, 147–148, 181, 187, 193, 195, 199
conserved current 82, 90–91, 92, 93, 94, 95, 96, 99, 171, 175–176, 186, 191, 193, 194
conserved property 48, 81, 95, 99, 101–102, 177, 186–187
conserved quantity 4, 93–94, 95, 97, 165, 171, 175–176, 191–194
contravariant 89, 141
corpuscle 111–112, 118, 176, 197–198
corpuscle-wave duality 4, 197
Coulomb charge 31, 36–37, 44–46, 99, 118, 162, 185–186, 189
Coulomb-type electric charge 49, 200
coupled field equations 92
coupled fields 128
coupling constant 86, 93, 117–118, 148, 164, 170, 179, 193
covariance 27, 34–37, 43–44, 46, 74–77, 84, 101, 111, 138, 143, 145, 157, 162, 189, 199
covariant derivative 85, 88–89, 186
curvature 89, 145, 172, 174

D field 6, 81, 83, 94–95, 96, 101, 104, 111, 119–121, 125–126, 128–131, 133–134, 152, 164, 171, 173–177, 180, 194–195
dion (ð) 23, 104–105, 110, 119–120, 130, 133, 137, 145–146, 177–179, 181, 183, 191, 193, 194–197
dion's mass 178
Dirac equation 51, 104, 119, 144, 147–148, 150, 152–155, 157–161, 163, 166, 173, 179–180
Dirac matrices 147, 150–151
dynamical symmetry breaking 119–120

effective velocity 132, 195–196
eigenfunction 54–55, 141, 143–144, 158–159
eigenvalue 4, 48, 54–55, 60, 66, 71, 96, 143–145, 158, 176
Einstein equation 25, 119, 144, 189
electric charge 18, 20, 28, 31, 33, 35–36, 43–47, 73, 81–82, 99–100, 102–105, 116–118, 147 149, 153–156, 160, 162–163, 165, 168, 170–171, 173–177, 180–183, 185, 189, 197–198
electric moment 163, 165, 168, 174, 179–180
electrodynamics 28, 37–39, 45, 73, 76, 82, 104, 189
electro-kinetic moment 163, 165–167, 169, 180

https://doi.org/10.1515/9783110713183-014

electromagnetic field 28, 34–37, 39, 43, 45, 47–48, 73–76, 84, 101–103, 115, 147, 151–155, 163, 169, 175, 177, 185, 196
electromagnetic interaction 43, 46–47, 48, 49, 76, 102, 113, 118, 147–148, 157, 162, 164, 168, 170, 174, 180, 182, 187, 199
electron 16–17, 19, 31–33, 47–48, 75, 84, 114–115, 117, 128, 148–149, 154–155, 163, 165–166, 176, 181, 198
electroweak field 128
electroweak interaction 99–101, 181–182, 199
equations of motion 26, 46, 76–77, 92, 95, 99, 175, 191
equivalence 7, 13, 15–17, 19–22, 24, 26–27, 35, 46, 138–139, 146, 150, 154, 156, 185, 189
equivalence principle 13, 21, 23–24, 26–27, 34, 36, 43, 45–46, 138–139, 162
Euler-Lagrange equations 92
exotic particles 6
extended Dirac equation / modified Dirac equation 70, 156–157, 159–162, 165–166, 180

fermion 5, 8, 17, 20, 24, 81–82, 89, 103, 120, 146, 191, 196, 198
Feynman diagram 135
field D 91–92, 96–97, 101–102, 104, 119–120, 126–131, 135, 155, 166, 168, 175–177, 195
field equation 28, 82, 86, 92, 95, 99, 121, 138, 141, 144, 158, 175, 191
field source 22, 30, 81, 103, 193
field tensor 171, 174
field charge 7, 11, 37, 39, 43–45, 46, 49, 51, 67, 71, 82, 92, 93–95, 97, 99–105, 111, 112, 115, 119–120, 122, 125, 128, 131, 135, 137, 143–145, 147, 157, 159, 161, 175–177, 183, 186, 188–192, 194–196, 199
field-charge current 51, 94, 112, 198
field-charge siblings 8, 82, 186
Finsler geometry 145, 147, 187
flavour 17–18, 20, 43, 45, 100, 102, 120, 126, 181–182, 197
flip-flop model 110–111, 191
four-current 36–37, 43, 49, 155, 192
four-momentum 30, 45–46, 49, 149–150, 192
four-potential 28, 37–38, 75
four-vector 37–38, 45, 88, 156, 185
free state 31, 33, 44, 99–100, 103, 109–110, 112–113, 117, 198
free-state 109
full electric moment 167–168, 170
full magnetic moment 165, 170
fundamental interaction 99, 101, 103, 119, 145, 161, 199

gauge field 19, 28, 43, 48, 75–77, 78, 82–86, 89–96, 99, 101–102, 104–105, 115, 119–121, 123, 134, 148, 154–155, 175–177, 179, 190–191, 193–194, 197–198
gauge invariance 28, 34, 76–77, 82, 84, 86, 191
General Theory of Relativity (GTR) 13, 21–25, 27, 45, 77, 119, 189
Goldstone boson 119–121, 126
gravitational equation 22, 139, 143, 145, 187
gravitational field 22, 25, 28, 36–37, 43–45, 74, 99, 101–102, 116, 138, 141, 146, 161, 178, 185
gravitational interaction 30, 43, 45, 102, 138, 146–147, 157, 197
gravitational mass 14, 21–30, 36, 43–45, 49, 73, 99, 113, 116–117, 138, 140, 143, 146, 186, 189, 200
group of hypersymmetry 57, 95, 97, 105, 139, 164
group of HySy 60, 119, 192

hadron 17, 179, 181–182
Hamiltonian 14, 28, 30–31, 34, 44–46, 74, 77–78, 85, 99, 103, 110–111, 113, 115, 120, 138, 143–145, 148–149, 151–152, 154–155, 157, 161, 165, 168, 170, 179–180, 186, 188, 190–191, 193, 198
harmonic oscillator 110–111, 118, 190
Higgs boson 4–5, 100, 185, 187
hypersymmetry, HySY 3, 6, 8–9, 33, 51, 54, 69–70, 81, 96, 101, 110, 119–121, 124–126, 128–130, 132–134, 137–139, 143, 145, 147, 152, 157, 161, 177, 183, 185–188, 190–193, 195–200
HySy group 124, 144, 157–158

identity 7, 13, 15–18, 19, 20, 21, 22, 27, 35, 46–47, 56–58, 60, 128–130, 133, 138, 151, 155, 185
inertial mass 8, 13–14, 15, 21–27, 29–30, 36–37, 44–45, 73, 99, 103, 113, 116–117, 138, 140, 142, 145–147, 149–150, 160, 185–186, 189

interaction 4, 7–8, 13, 22, 24, 28–29, 31–33, 43–44, 46, 48, 51, 75–76, 81–82, 85–86, 95, 97, 99–105, 107, 109–110, 112–121, 125, 132, 134–139, 141, 143, 145–147, 154–155, 159, 161–163, 165–166, 168–171, 175–178, 180–182, 185–191, 194–195, 197–199
intermediate boson 7, 111, 119–120, 126, 134–136, 174, 179
intermediate model 109, 112–113
invariance 4, 7, 15–16, 19, 24, 26, 28, 34, 37, 43, 48, 51, 75–78, 84–85, 88, 92, 95, 101, 104–105, 115–116, 119, 121, 124, 134, 137–139, 144, 148–150, 154, 159, 162–163, 175, 177, 181, 187, 193–194, 197, 199–200
isospin, isotopic spin 6, 8, 19, 20, 47, 48, 76, 79,93, 96, 97, 99, 101–105, 109–111, 115, 117, 119, 121, 124, 127, 131, 133–135, 143–145, 147, 152, 154–155, 157–162, 165, 175, 176, 177, 181, 183, 186–187, 191–194, 199
isotopic field-charge field 115, 121, 195
isotopic field-charge, IFC $(\daleth_v, \daleth_T)$ 41, 43, 45–47, 49, 51, 57, 65, 73, 74, 94–97, 99–101, 107, 109, 111, 112, 114–116, 120–121, 126, 128, 131, 133, 138, 147–149, 152, 160–162, 164, 176–177, 181, 185, 188, 193–195, 197–199
isotopic field-charge state 73, 110–113, 115–117, 183, 194, 198
isotopic state 7–8, 44–45, 47–49, 51, 71, 81–82, 95–97, 100–101, 103, 109, 116, 143–144, 159, 175–176, 181, 185, 190, 194, 197

kinetic gauge field 90, 96–97, 104, 137, 143–144, 148, 155–158, 160, 163, 165, 168–170, 175–176, 194, 199
kinetic potential 171, 174
kinetic state 96, 100, 103, 110–114, 117, 137, 143–144, 146, 149, 157–158, 176, 181, 183, 190, 194, 198

Lagrangian 25–26, 28–30, 44, 73, 75–77, 84–85, 89, 92, 95, 99–100, 103, 135, 146, 154–155, 157, 168, 174–175, 180, 186, 189, 197
Lagrangian density 85–86, 89, 92, 99
lepton 17, 43, 100, 102–103, 181, 194
limit velocity (of HySy) 130, 134
localisation 39, 73–74, 77, 83–84, 190, 198
Lorentz charge 185–186, 189
Lorentz group 19, 197, 199
Lorentz invariance, Lorentz invariant 24, 37, 39, 73, 84, 86, 95, 134, 138–139, 148–149, 161–162, 175, 177, 189, 193, 197, 199
Lorentz transformation 14, 22–23, 25–26, 37, 39, 68–70, 73, 77, 84, 88, 95, 111, 119, 121–125, 128, 143, 148, 150, 157, 174–175, 193, 197, 199
Lorentz type electric charge 176, 200

magnetic moment 44, 99, 166, 170
magneto-kinetic moment 163, 165, 169, 174, 180
mass 133
mass of dions 119
mass of gravity 7, 19, 43–45, 117–118
mass of inertia 7, 19, 44–45, 116–118
massive bosons 194
matrix algebra 51, 57, 144, 158
matrix genetics 66
meson 17, 104, 183
Møller scattering 33, 45, 114, 128

neutron 17, 20, 46–48
New Physics, NP 3–5, 6, 81, 185, 188, 196, 200–201
Noether current 82, 88–89, 90, 91, 92, 97, 176, 186, 189, 193, 196
Noether's theorems 34
non-Abelian 34, 82, 84, 90, 96, 101, 104, 176

object 13, 16–17, 19–21, 27, 36, 44–46, 74, 101, 104, 110–111, 113, 116–118, 183, 188, 197
operator 4, 54–55, 60–61, 65–66, 71, 85, 89, 93–94, 96–97, 143–145, 151–153, 158–160, 162–163, 170, 176–177, 194

Pauli matrices 57–61, 69
phase space 94, 175, 190, 199
physical property 37, 101, 104, 177
potential 74
potential state 96, 105, 110–112, 118, 143, 146, 149–150, 157, 169, 176–177, 181, 183, 190, 194
precession 123–124, 127, 131–134, 152
precession angle 129–130, 195
probabilistic model 110

property 16, 18–19, 20, 21, 27–28, 36, 47–48, 58, 65, 71, 74, 81–82, 95–97, 99–103, 114, 117, 137–138, 143–144, 155, 157, 176, 186, 190–191, 194, 198–199
proton 17, 20, 46–48, 117, 178, 183
pseudo-unitary 59, 61, 64

quanta of the D field 82, 95, 97, 103–104, 134, 175–177, 194
quantisation 3, 175
quantum chromodynamics, QCD 21, 115
quantum electrodynamics, QED 28, 31, 33, 38–39, 46, 56, 66, 68, 70, 109, 113, 119, 147–149, 152–154, 163, 170–171, 173, 179, 187, 189
quark 17–18, 19, 20, 21, 43, 47, 99–100, 102–103, 110, 126, 181–183
quaternion 63, 104, 159
quaternion algebra 51–52, 54, 62

real velocity 131–132, 196
relativistic 5–6, 13–14, 25, 34, 73, 84, 92–93, 117–119, 134, 143, 149–153, 157, 168–169, 177, 180, 189, 196
relativistic covariant fields 152
renormalised 194
representation 53–55, 57, 59–61, 63–69, 74, 85, 93, 96, 124, 139, 176, 187, 192
Ricci tensor 145–146
Riemann geometry 145
rotation angle 123, 126–127, 130, 195
runaway motion 154, 198

scalar field 51, 84, 92, 111–112, 114, 148, 151, 176, 190
scalar potential 28, 32–33, 36, 39, 43–44, 49, 61, 99, 109, 111–112, 151, 156, 192
scattering matrix 33, 112, 185
Schrödinger equation 168, 179–180
self-adjoint matrices 57
self-adjoint physical operator 58
simplicity principle 188
SO(3,1) 85, 93
Special Theory of Relativity (STR) 24, 73, 77
spin 16–17, 20, 48, 71, 96–97, 101, 114–115, 119, 143, 146, 152, 157–158, 168, 170–171, 180–181, 187, 192
spin-0 field 119
spinor 54–55, 61, 64–65, 68, 96
spins 20
spontaneous symmetry breaking 121, 127, 129–130, 133, 194–195
spontaneous symmetry breaking angle 130
Standard Model, SM 3–4, 5, 6, 7, 8, 18, 21, 23, 38, 43, 47–48, 81, 91, 93, 95, 101–105, 111–112, 119–121, 125–126, 128–130, 133–134, 136–137, 185, 188, 195, 200
state function 4, 48, 54–55, 144–145, 152, 157–159
stress-energy tensor 22, 25–26, 30, 39, 45, 138, 139
stress–energy tensor 138–141, 145, 157
string theories 5–8, 139, 185, 188
strong field 43, 99–101
strong interaction 8, 47–48, 102–103, 118, 179, 183, 189, 195–196, 199
SU(2) 48, 60–62, 63, 71, 74, 85, 93, 95–97, 100–101, 105, 134, 138, 145, 162, 164, 174, 176, 187, 192, 199–200
SU(3) 19–20, 74, 85, 93, 199
supersymmetry, SUSY 3, 5–7, 8, 9, 51, 81, 137, 185, 187–188, 195, 200
symmetry 4–5, 6, 15–16, 19–20, 26, 28, 30, 32–33, 39, 47–48, 68, 70, 74–77, 81, 85, 93–94, 96, 100–101, 103–104, 114, 116–117, 119–120, 128, 135, 138–140, 145, 148, 155, 157, 162, 164, 176–177, 181, 186, 188, 191, 199
symmetry breaking 114, 120
symmetry group 24, 26, 74, 85, 93–94, 96, 120, 174, 176, 197, 199
symmetry principle 4–5, 6, 32–33, 75–76, 81, 188

transformation 14–15, 16, 17, 18, 19, 22–26, 28, 30, 34, 36–37, 39, 43–45, 48–49, 51, 55, 58, 65–66, 68–70, 73–75, 77–78, 81, 84–86, 89–90, 92–96, 99, 101–102, 104, 113, 119–121, 123–130, 133–135, 138–142, 144–145, 148, 151–152, 155–156, 158, 160–162, 164, 175–176, 186–187, 189, 192–195, 198–199
transformation group 55, 84, 96, 176, 192, 193

U(1) 74, 85, 93, 100, 162, 199

vector algebra 51–53
vector field 43, 51, 82, 84, 92, 111–112, 115, 151, 154, 190
vector potential 31–33, 36, 38, 109, 113, 151, 156, 163–164, 167, 170
velocity space 73, 77, 82–83, 88, 94, 175, 190, 192
velocity-dependent field 8, 25, 74–75, 84, 119, 122–124, 134, 145, 166, 186, 189, 192
velocity-dependent phenomena 73, 82
vierbein 46, 55, 141–142, 155, 168
violations of symmetry 4
virtual "coupling" spin 170–171, 180–181

wave 4, 23–24, 27, 111–112, 117, 149, 176, 197–198
wave equation 149, 151–153, 168
wave function 4, 34, 96, 110–111, 115, 117, 143, 149, 153–154, 157, 176, 190, 198
weak charge 100, 102–103, 182–183
weak field 43, 74, 99–100
weak hypercharge 100, 103
weak interaction 47, 102–103, 105, 128, 174, 179, 181–182
weak isospin, weak isotopic spin 48–49, 97, 100–101, 103, 187, 192
weak mixing angle 128
Weinberg angle 126, 128–129
Wheeler-deWitt equation, WdW 143, 145
Wigner–Thomas rotation 123
Wigner's little group 114

Yang-Mills gauge 48
YM field 83, 96, 101–102, 104, 115
YM theory 89

Δ exchange 146, 162, 181, 183
τ algebra 49, 54–57, 59, 64–66, 69–71, 95, 105, 124, 138–139, 144, 160, 161, 187, 192, 200

www.ingramcontent.com/pod-product-compliance
Lightning Source LLC
Chambersburg PA
CBHW061413210326
41598CB00035B/6200
9783110713176